Mathematisation and Demathematisation

Mathematisation and Demathematisation
Social, Philosophical and Educational Ramifications

Edited by

Uwe Gellert
Universität Hamburg, Germany

Eva Jablonka
Freie Universität Berlin, Germany

SENSE PUBLISHERS
ROTTERDAM / TAIPEI

A C.I.P. record for this book is available from the Library of Congress.

Paperback ISBN 90-8790-011-2
Hardback   ISBN 90-8790-012-0

Published by: Sense Publishers,
P.O. Box 21858, 3001 AW Rotterdam, The Netherlands
http://www.sensepublishers.com

Printed on acid-free paper

DEDICATION

Festschrift is a German word used to denote a volume of writings by various authors which is presented as a tribute, nowadays usually to a prominent scholar.

The first Festchrift was published in 1640 by Gregor Ritzsch in Leipzig on the occasion of the bicentenary of the invention of the printing press ["Jubilaeum Typographorum oder zweyhundertjähriges Buchdrucker-Jubelfest"].

Poets from all over Germany contributed.

The presentation of such a Festschrift has become a tradition in German-speaking academia and is made by fellow scholars as a token of their affection and respect, and is considered a great honour.

We are pleased to be able to continue this tradition, although in this case, as the title might suggest, the contributions can hardly be considered poetic.

This book is dedicated to a unique personality, and is the occasion of our being able to express our warm gratitude for all she has done for us: her ability to fire our imaginations, and through stimulating discussions show us that, without a transdisciplinary approach, the crucial issues cannot be understood, and that research can only be relevant if it is sustained and nourished by a vision. She is a scholar who did not allow us to forget that mathematics education should not be separated from its political and social context. Her introducing us to the international scientific community saved us from the fate of becoming too deeply rooted in the German community of mathematics educators.

This book is dedicated to Professor Christine Keitel on the occasion of her birthday.

Uwe Gellert and Eva Jablonka (January, 2007)

CONTENTS

EVA JABLONKA AND UWE GELLERT

MATHEMATISATION – DEMATHEMATISATION

This introductory chapter attempts to discuss and thereby clarify the meanings of *mathematisation* and *demathematisation*. Given the sufficiently cryptic character of the title of this volume, such a clarification seems particularly important.

The title of this volume is motivated by our belief that the notions of mathematisation and demathematisation describe two phenomena, which are dialectically related. In our view, the power of the concept of mathematisation in terms of its heuristic value and theoretical fruitfulness might be considerably increased if the reverse process – demathematisation – were conceptually integrated.

Scholars from the diverse strands of research have made use of the notion of mathematisation in order to coin a process in which something is being rendered more mathematical than it has been before. There is a tradition of discussing mathematisation as a didactic principle, for example in the context of *Realistic Mathematics Education* in the Netherlands. On the other hand, some use the concept of mathematisation for the description and analysis of the social, economical or political processes in which relationships between the participants become increasingly formal. Because these apparently different conceptual understandings of mathematisation are used in parallel within the field of mathematics education, it seems appropriate to first point out conceptual differences.

A common concern of all authors of this volume is the issue of social availability of mathematical knowledge. Reference is made more or less explicitly to the educational dimension of processes of mathematisation and demathematisation. Together, the contributions reveal a rather complex picture: They draw attention to the importance of clarifying epistemological, societal and ideological issues as a prerequisite for a discussion of curriculum. Such a view implies a critique of those curricular conceptions focussing on mathematisation, which lack adequate forethought. In line with this view, in this chapter, after elucidating the concepts of "horizontal and vertical mathematisation" and "mathematisation in the course of mathematical modelling", we discuss in more detail the work of scholars who concentrate on analysing mathematisation and demathematisation as a social process.

U. Gellert, E. Jablonka (eds.), Mathematisation – Demathematisation: Social, Philosophical and Educational Ramifications, 1–18. © *2007 Sense Publishers. All rights reserved.*

MATHEMATISATION AS A DIDACTIC PRINCIPLE

It is in fact a very old tradition to use a simple description of an everyday or professional activity as a paradigmatic representation of a class of similar problems for the purpose of introducing a mathematical method for solving them. These mathematical tasks contain no redundant information, no data is missing, the answer is well defined and the results are never brought into operation. The term mathematisation is used by some to describe the activity of the students who are dealing with these types of word problems. The object, which is to be mathematised, need not be as simple as a traditional textbook problem, but can also consist of a description of a more or less authentic out-of-school situation, which by its alleged authenticity is obviously more complex than a simple textbook problem.

Horizontal and vertical mathematisation

Mathematics lessons often consist of students being introduced and accustomed to read and solve textbook problems, many of which are contextualised. One of the reasons for the difficulties teachers and students, who are faced with "realistic" problems, experience, is that school mathematics, if about optimisation when shopping, for example, is neither mathematics nor shopping.

Dowling (1996, 1998) describes the effects of this didactic practice as *the myth of participation* and *the myth of reference*. He identifies two distinct messages in school mathematics texts. One, the "myth of participation", suggests that mathematics deals with the public domain. Realistic problem situations in the form of narratives from the perspective of a person acting in a practical situation are used as an introduction into the esoteric domain of mathematics. This strategy supports the view that everyday practical knowledge is constitutive of mathematical knowledge. But it is likely that this knowledge turns out to be an insufficient base for solving the contextualised problems because the mathematical solution has a structure that is different from any practical solution. The second message is the "myth of reference". It is conveyed through problem settings that are constructed mathematically and only retain a trace of non-mathematical significations. It does not remain possible for the learner to evaluate the solution of the problems from a practical point of view. However, the 'recipients of this message' are expected to believe that behaving mathematically in such situations would be in their own interests. According to Dowling, both of these messages are myths: The application of mathematics in practical (domestic) activities conceals its generalisability, and the expression of domestic activities in mathematical terms conceals the cultural arbitrariness of the mathematical principles projected into them.

In both cases, students need to transcend the everyday context, in which a problem is formulated, by converting the semantic structure of the text into a mathematical description. Of course, since problems given in textbooks generally do not claim to mirror problematic situations authentically, the mathematisation required from the students is essentially an artificial activity.

According to Treffers (1987) the process of formulating a mathematical description might be called a *horizontal* mathematisation. The semantic transfer occurs horizontally between different domains of description: from the public to the esoteric. In the mathematics classroom, particularly in the lower grades, it is typical to switch between these two domains. First mathematical concepts, like the cardinal and ordinal aspect of numbers, are built from activities or relationships within the public sphere. These concepts, as Freudenthal (1983) points out, help to organise many everyday phenomena. Also in the higher grades, mathematical meaning is often developed from out-of-school experience.

Adler (2001) provides telling empirical accounts of secondary mathematics teachers being confronted with this changing between the public and esoteric language of description. A dilemma emerges: If classroom talk concentrates on the public language of description in order to help students construct meaning, then the mathematical knowledge of students tends to remain in the public domain of its origin. If, on the other hand, the classroom talk is mainly esoteric, then the individual construction of meaning appears to be more difficult. A similar point is made by Vithal and Skovsmose (1997) in respect to ethnomathematics. They argue that career aspirations of students and parents would be ignored if mathematics education concentrated too long on local knowledge rather than providing the opportunity for students to enter the esoteric domain of academic mathematics.

However, it can be argued theoretically that the switching itself is intrinsically problematic, because everyday knowledge is often contradictory across contexts. According to Bernstein (1996, 1999), everyday knowledge, or common-sense knowledge, is context-dependent, segmentally organised and consistent within each segment, but segments overlap and knowledge organisations often do not match. Bernstein calls the discourse referring to this type of knowledge a *horizontal* discourse, compared to a *vertical* discourse, which takes the form of a systematically principled and hierarchically organised structure, such as in the sciences, or of a specialised language, as in the social sciences and humanities. Horizontal mathematisation as used by Treffers (1987) can be described, within Bernstein's theoretical framework, as a transfer between horizontal and vertical discourse. It is problematic that horizontal knowledge remains contradictory across its segments, whereas vertical knowledge strives for coherence and a principled structure. In the process of horizontal mathematisation, the contradictoriness of the horizontal discourse is mostly ignored. The fiction is, that abstraction from extra-mathematical contexts to mathematical concepts and structures is possible and straightforward, but, actually, this process is a step from the contradictory world to a coherently organised esoteric sphere that has long since cut off its everyday roots.

Once the classroom activities occur within the esoteric domain of academic mathematics, e.g. defining concepts and proofing theorems, students get introduced into vertical discourse. This is what Treffers (1987) calls a *vertical* mathematisation. It is essentially an introduction into the organised structure of mathematical knowledge. Thus, Treffers's separation of vertical and horizontal mathematisation is reminiscent of Bernstein's discourses.

Horizontal and vertical mathematisation – in the curricular conception laid out by Treffers (1987, 1991) and colleagues (de Lange, 1996; Gravemeijer, 1994; Streefland, 1991; Treffers & Goffree, 1985) – define mathematisation as a multi-step activity for students, aiming at the exploration of mathematical structures. The focus is on the mathematics, not on the 'realistic' situations from which the mathematisation is hoped to be derived. Everyday situations are valued mostly for the alleged motivation and illustration that they invoke. In short, this kind of mathematisation is concerned with its ends (which reflect the curriculum) rather than with the process itself.

Mathematisation as a didactic principle is often underpinned by the assumption that the historical development of basic mathematical concepts can be used as a guideline for the learning process of an individual:

> It is a characteristic of mathematical activity, according to the realistic view, that the build up of elementary skills can take place via a process of reinvention or independent construction. (Treffers, 1991: 31)

The "realistic view" is a rudiment of the genetic principle in epistemology as formulated by developmental psychology. This principle postulates parallelism of ontogenesis and historiogenesis. Damerow, in this volume, develops a framework for a historical epistemology of the concept of number in order to answer the question of whether in the history of mathematical knowledge stages of development can be identified which at the same time constitute and constrain individual cognition. Arithmetical thought is reconstructed as an outcome of the material culture of calculation.

From 'real world models' to mathematical models

The process of translation from 'reality' to mathematics and back is in the focus of curricular conceptions figuring under the title *mathematical modelling* (de Lange, Keitel, Huntley, & Niss, 1993; Houston, Blum, Huntley, & Neill, 1997; Matos, Blum, Houston, & Carreira, 2001). In the mathematical modelling perspective, the teaching and learning of mathematics is organised along a didactically simplified view of applied mathematics. The fundamental idea of this approach is that students should be introduced to, and involved in, what are seen as the core sub-processes that applied research runs through when tackling authentic social or technical problems.

In order to give a structured description of 'real world problem solving', the 'world' is separated into two spheres: 'reality' and 'mathematics'. A 'real world problem', of course, is located in a 'real world situation'. According to the modelling perspective, the first step in coming to terms with the 'real world problem' consists of a problem description based on everyday knowledge of the situation. This description, in theory, is non-mathematical. It is regarded as a 'real world model' of a 'real world situation' in which the criteria for acceptable solutions are made explicit. Take as an example the 'real world situation' of some film enthusiasts discussing the scariest movie ever seen ('real world problem')[1]. After a lengthy debate they might agree on a set of criteria, which a movie should meet in order to be scary: Suspense is important, a good dose of realism as well, an emerging feeling of subjection, and some level of gore are all supposed to produce the best effects. These criteria make for the non-mathematical 'real world model'.

When the 'real world model' is converted into a formal description, e.g. a mathematical equation, the sphere of 'reality' has been left, and a 'mathematical model' is created. The mathematical description of the 'real world model' is the result of a translation from 'reality' to 'mathematics'. This 'translation' into the 'mathematical model' is by no means straightforward. It often includes, firstly, quantifying different non-mathematical characteristics and, secondly, relating these to each other mathematically. In the example above, it is not clear from the outset, how *a good dose of* realism and *some level of* gore might be best expressed mathematically, and whether they are equally important. Within the curricular conception of mathematical modelling, the process of 'translation' from the 'real world model' to the mathematical model is called *mathematisation*.

The formal problem description allows for mathematical concepts and algorithms to produce mathematical results. According to this simplified description of modelling, these results need to be tested against the 'real world situation' (or its 'real world model') from which the complex process started. Since the results often do not match the problems, reformulations of the 'real world model' and modified mathematisations might be necessary. As a consequence mathematical modelling is described as a circular process. Since within a classroom activity the results are never put into operation there is no real problem of validation.

It has been argued elsewhere (Gellert, Jablonka, & Keitel, 2001), that the epistemological underpinnings of applied mathematics are not adequately represented in the conception of mathematical modelling as described above. The symbolic technology at hand, as well as measuring devices, and domain-specific constraints and theories, all influence the form a 'real world model' will take. So the 'real world model' is by no means a description of the 'real world problem' (Cooper & Dunne, 2000; Gates & Vistro-Yu, 2003; Jablonka, 1997).

In the mathematics classroom, the fiction that 'real world models' are independent of the mathematical technology at hand is even harder to maintain: How can students construct non-mathematical 'real world models' according to their interests in distinct aspects of the situation and at the same time be aware of the fact that they are sitting in a *mathematics* lesson, and that they will have to mathematise and solve the 'real world models' they are building by means of their (limited) mathematical knowledge? The models students construct tend to relate to the mathematical tools they have developed shortly beforehand.

Mathematisation within the circular process of mathematical modelling is – epistemologically regarded – a potentially misleading construct and it is – pedagogically – of debatable value. On the one hand, the circular model of mathematical modelling adequately acknowledges the contingencies of problem definition and formalisation; on the other hand, it tends to obscure the informative power of mathematics. Mathematics is not only the sphere where formalised problems find their solutions; mathematics is from the outset the vantage point from which the problems are construed.

MATHEMATISATION AS A SOCIAL PROCESS

Mathematisation, in the context of mathematical modelling as a didactic principle, falls short of grasping the fact that the 'world' students live in is already interspersed with constructions and processes based on mathematics. Mathematics is a means for the generation of new realities not only by providing descriptions of 'real world situations', but also by colonising, permeating and transforming reality. Models become the reality, which they set out to model. Consequently, any discussion of mathematisation has to take into account the social process by which mathematical models are developed, implemented, accepted, and obscured.

Mathematics shaping society

According to Davis and Hersh (1986), we are living in a Descartesian world. Mathematics has penetrated many if not most parts of our lives. It has capitalised on its abstract consideration of number, space, time, pattern, structure and its deductive course of argument, thus gaining an enormous descriptive, predictive and prescriptive power. Within science, there is hardly any theory, which is not formulated in mathematical terms. In sociology, psychology and education, quantitative studies are highly valued. It is hardly impossible to understand any theory of economics without a solid mathematical background. In all these fields of human activity mathematics can be regarded as the grammar of the particular scientific discourse and as a universal tool. However, mathematics being the grammar or the tool implies that the characteristics of this grammar strongly influence the development of the fields in which the use of mathematics is made. It turns out to be difficult, if not impossible, to integrate any idea that cannot be formulated in mathematical terms into an accepted body of mathematically formulated theories.

The impact of mathematics is by no means restricted to scientific activity. Mathematics-based decisions affect the social interactions in technological societies on many levels (for a long list of examples see Davis & Hersh, 1986: 120-121). On the level of the national distribution of state salaries, pensions, and social benefits, political decisions often are made and communicated on the basis of formulae and diagrams, which themselves rely on mathematical extrapolations of demographic and economic data provided by experts. On the level of interpersonal relations, mathematics-based communication technologies have already changed the habits and styles of private conversations. Of course, the mathematics often is invisible, as in mobile phones and Internet chat forums, or it is just recognised on the surface as a medium of presentation.

By drawing attention to the many ways in which mathematics shapes society and exerts considerable influence on our everyday lives, Davis and Hersh (1986) introduce the concept of mathematisation, denoting both the process and the product of ongoing formalisation. Davis (1989) focuses on another crucial aspect of mathematisation that has not yet received much attention: Is mathematisation a naturally occurring phenomenon or, if not, who is in charge, and whose intentions become realised?

Living with realised abstraction

Mathematics, when creating and using abstract concepts, presupposes abstraction. The basic mathematical abstractions (particularly such as the number system) are not only abstractions in the mind, but also in action. As Sohn-Rethel (1978) argues, the abstraction of the exchange value of goods is performed in trading situations, as is the abstraction of a general quantified measure for goods which is independent from their special qualities. He calls an abstraction, that is not thought but articulated in a social action, a *real abstraction*. Mathematics, as Freudenthal (1983) points to, has developed over the centuries by reflecting on these real abstractions – a process that could be reversed for educational purposes in form of a didactical phenomenology. Such a reflection on real abstractions happens in the mind as a *thinking abstraction* (Sohn-Rethel, 1978). Since thinking abstractions do not take place in the action of people, but in their imagination, they are extremely flexible. Mathematical thinking has the power of hypothetical reasoning: It is possible to calculate some consequences of actions before these are carried out. Nobody needs to be afraid of the immediate consequences of any thinking abstraction.

In the long run, however, the world of thinking abstractions transforms back to what Keitel, Kotzmann and Skovsmose (1993) describe as a system of *implicit* knowledge. In many cases, we are neither aware of the circumstances under which a particular thinking abstraction has been processed, nor of the purposes for their initiation. The social origins and the history of many of the mathematisations Davis and Hersh (1986) describe are immersed. Technology, including social technology, functions as a black box – and the constitutive abstraction needs not to be reflected upon anymore. The substitution of abstraction processes by black boxes produces what Keitel, Kotzmann and Skovsmose (1993) call *implicit mathematics*.

In order to stress the point that thinking abstractions – after having been derived from real abstractions and developed by hypothetical trial and error – shape the technology with the help of which we organise much of our life, Keitel, Kotzmann and Skovsmose (1993) introduce the notion of *realised abstraction*. Thinking abstractions become materialised, they become part of our reality, and most of the time we do not ask where they come from or what they are – there is no necessity for doing so. Our time-space-money-system is a striking example for the implicitness of the underlying abstraction processes. We are dealing with our system of space, time and money as if no other ways were possible. Indeed, these coordinates of our life seem natural to us. Where people share other conceptions

7

about these (e.g., Harris, 1991) – while living in the same 'global village' – their constructions appear virtually incomprehensible to us.

PROCESS AND EFFECTS OF MATHEMATISATION/ DEMATHEMATISATION

The concept of realised abstraction may make us understand that the mathematisation of our world is just one side of the coin. On the other side, most people need to carry out less mathematics explicitly. Technology is more effective. The existence of materialised mathematics in the form of black boxes reduces the importance of mathematical skills and knowledge for the individual's professional and social life. A *demathematisation* process is taking place (Keitel, 1989; Keitel, Kotzmann, & Skovsmose, 1993):

> This term [demathematisation] also refers to the *trivialisation* and *devaluation* which accompany the development of materialized mathematics: mathematical skills and knowledge acquired in schools and which in former time served as a prerequisite of vocation and daily life lose their importance, and become superfluous as machines better execute most of these mathematical operations. (Keitel, Kotzmann, & Skovsmose, 1993: 251)

The process of demathematisation affects strongly the values associated with different kinds of knowledge and skills. For the user of technology it becomes more important to, first of all, simply trust the black box and, then, to know when and how to use it – for whatever purpose. If it turns out to be inefficient, ineffective, erroneous or disastrous, nobody can be blamed – it was the technology's fault. From this perspective, demathematisation reduces the feeling for, and the acceptance of, responsibility: a car's antilock braking system is taken as a licence for driving fast; formal assessment of personal creditworthiness ensures that a loan is not approved to the wrong people. When faults still occur, then this is a request for a better technology – technology that is designed to be foolproof.

Curiously, demathematisation is also connected with the myth of the infallibility of technology. When airplanes crash or nuclear power stations run into problems, it is often attributed to human error. However, safety precautions and regulations of such risk technologies are the product of balancing the probability and impact of accidents against the costs of preventive measures. Of course, this is again a matter of realised abstraction.

Chevallard (1989; reprinted in this volume) draws attention to the importance of a process he describes as follows:

> Implicit mathematics are formerly explicit mathematics that have become "embodied", "crystallized" or "frozen" in objects of all kinds – mathematical and non-mathematical, material and non-material –, for the production of which they have been used and "consumed". (Chevallard, 1989: 50)

The theory of how sophisticated mathematical techniques, that were developed by "mathematical workers", become simplified by algorithmisation and crystallized in technological tools, is reminiscent of Karl Marx's observations on machinery:

> The machine proper is therefore a mechanism that, after being set in motion, performs with its tools the same operations that were formerly done by the workman with similar tools. Whether the motive power is derived from man, or from some other machine, makes no difference in this respect. From the moment that the tool proper is taken from man, and fitted into a mechanism, a machine takes the place of a mere implement. The difference strikes one at once, even in those cases where man himself continues to be the prime mover. (Marx, 1961: 374).

Chevallard (1989; reprinted in this volume) describes the dialectic between implicit and explicit mathematics:

> The greatest achievement of mathematics, one which is immediately geared to their intrinsic progress, can paradoxically be seen in the never-ending, two-fold process of (explicit) demathematising of social practices and (implicit) mathematising of socially produced objects and techniques. (Chevallard, 1989: 52)

It is indeed this "paradox", which has to be the starting point of any discussion of the value of mathematical skills for an individual (e.g. Coben, 2003: 47-53; FitzSimons, 2002a, 2002b; Jablonka, 2003). As issues of curriculum became increasingly important with the development of mathematical technology, some scholars from the field of mathematics education have continued to explore the relationship between mathematisation and demathematisation. This ongoing discussion includes issues of the relationship between mathematics, society and technology; of a shift of values linked to mathematical knowledge; and of the importance of a democratic competence within demathematised societies.

The following discussion does not intend to be a comprehensive literature review, but to provide a base for further exploration by concentrating on some essential work focussing on those issues.

Mathematisation/ demathematisation through technology

Drawing on Mumford (1967), Keitel (1989) illustrates the role and possible effects of technology by the example of the mechanical clock[2]. The construction of the clock is based on the perception of the movement of the planetary system:

> This approach is generalized and condensed to a mathematical model, transformed into a technological structure, and as such installed outside its original limited realm of significance. Earlier human perceptions of time, which had grown out of both individual and collective experiences and remained bound and restricted to these, were now rivalled and ultimately substituted for by the novel way of perceiving time. (Keitel, 1989: 9)

9

The first effect of this technology is a mathematisation that makes it possible to measure time precisely and independently from the quality of the processes measured. The abstraction of comparability is presupposed. The meaning of time passing slowly in some instances, and running more quickly in others, can then only be understood metaphorically. The objective character of the mechanical clock denies subjective experience of time. The specific (subjective) situation, in which time is measured, has lost its relevance. A formalisation has taken place. Time is no longer valid as a concrete sensory experience (cf. Keitel, 1989).

This objectification and formalisation still has tremendous implications: Time is regarded as the sum of arbitrarily regular units. The idea of the linearity of time is associated with the concept of progress and endless evolution. Mathematics as the grammar of science is reinforced:

> The mechanical clock extends the domain of quantification and measurability. Applying measure and number to time means measuring and quantifying all other areas, in particular those where time and space relate to one another. The measurability of time pushes forward the development of the natural sciences as (empirical) sciences of measurement (and hence objective sciences) and mathematics as the theory of measurement. (Keitel, 1989: 9)

Equally important, mathematics serves as the grammar of social order and social coordination. Keitel, Kotzmann and Skovsmose (1993) refer to F.W. Taylor's (1947) introduction of "scientific management": Every complex work process can be broken down into elementary components; the time necessary to carry out these elementary components can be measured; the time in which a complex work process should be finished is the sum of the small but many 'pieces' of time needed for the elementary components. Here, the measurement of time "objectively" determines work organisation.

Keitel sums up:

> Thus, the mechanical clock changed the relation between mankind and reality far beyond its original domain of application. It initiated the creation of a second nature totally reconstructing the first, exclusively admitting objective, mathematical laws, devaluing the authority of individual and collective (subjective) experience or insight. (Keitel, 1989: 9)

The "second nature" has replaced the first one and it appears as if mathematically conceptualised time were the most natural thing in the world. Indeed, nobody has to actually think about the consequences of this technological development or about the losses that its introduction has effected, when using a clock. The mathematical abstraction, which is encapsulated in it, has vanished from the surface – but it nevertheless continues to be effective:

> *Implicit mathematics* makes mathematics disappear from ordinary social practice. (Keitel, 1989: 10)

Technology can be characterised by its effect of making the underlying (mathematical) abstraction processes invisible. At the same time, technology facilitates the use of mathematics in social or technical situations precisely by liberating the user from the details of the mathematics involved. A curious correlation can be observed: Whereas the flexibility and potential of mathematical thought lies in its harmlessness – there is no immediate threat of changing the physical world by carrying out mathematical abstractions and calculations, "we can handle hypothetical states of affairs" (Keitel, Kotzmann, & Skovsmose, 1993: 250) – the materialised mathematics of technologies has lost its innocence.

The abstraction encapsulated in many technological artefacts brings about flexibility in their range of application. It is exactly this flexibility that makes them suspicious. Technology that has been developed with the help of mathematical thought for a particular purpose, is often "flexibly" applied to new situations where new mathematical thinking might have resulted in a qualitatively different technological solution if a new and different mathematisation had been introduced. Consequently, the effect of this transfer of technology is restricting the scope of problem solutions that could have been imagined. Thus, a restrictive effect can be attributed to technology (as realised abstraction), whereas the mathematics (as thinking abstraction) offers hypothetical explorations into new problem solutions.

Mathematisation/ demathematisation as a shift of values

Fischer (1993) draws on the sociologist Tenbruck's (1975) distinction of the *value of meaning* and the *value of utilisation* of scientific knowledge. The value of utilisation refers to the instrumental facet of scientific propositions, laws and procedures. These can be used efficiently as a means for solving social or technological problems. The value of meaning is associated with people's need and desire to understand human existence and to gain orientation for life in society. As Fischer points out:

> Value of meaning and value of utilization are not solely determined by the content of a scientific proposition. They depend on societal conditions, on already existing knowledge, and so forth. (Fischer, 1993: 115)

According to Tenbruck (1975), over time, scientific knowledge is subject to a shift of value. In the moment of its invention, scientific knowledge is valued mostly for its explanatory power. It provides new orientation for thinking about human existence – but it rarely has instrumental value from the outset. After some time, as Tenbruck argues, the value of meaning decreases: the new knowledge is built into the common explanatory framework. It becomes a matter of fact. This process is accompanied by an increase in its value of utilisation, the instrumental potential of the new knowledge is explored.

Fischer's (1993) interpretation of Tenbruck's thesis focuses on the development of mathematical knowledge. The value of meaning of mathematical knowledge was presumably at its height during the late 18th and 19th century when science and mathematics developed into separated organisations of knowledge with their own respective fields of enquiry. At that time, mathematics was increasingly viewed as going beyond science, as a universal method for explaining the world. It stimulated a new approach to language and logic and to describing social behaviour by means of mathematical concepts and laws. In short, during the period of the foundation of 20th century mathematics, its value of meaning had substantially increased.

Nowadays, in contrast, Albert Einstein's well-known saying that the closer a mathematical statement is to reality, the less certain it is, and that every reliable mathematical statement is far from reality, has become almost commonplace. A relativistic perspective is prevalent: we know that the connections between abstract and "sterile" mathematics and complex reality are not very tight. As Fischer (1993) puts it:

> One does not expect that mathematically invented propositions are fully correct in reality. ... Mathematics is often not used to describe reality, but to *construct a new reality*. For this it is excellent. (Fischer, 1993: 118)

The value of utilisation has replaced the value of meaning of mathematics. In order to illustrate that claim, Fischer (1993) discusses the highly interesting and relevant example of economic and financial policy. It does not make sense to take mathematical descriptions of trade balances, tax systems, stockbroking etc. as more or less adequate reflections of the financial reality. They actually construct this reality by setting up mathematical models that make rational financial decisions possible. Within this mathematically constructed reality, mathematics is automatically the best means for refinement of the constructions:

> We have a *circularity*: The more mathematics is used to construct a reality, the better it can be applied to describe and handle exactly that reality. (Fischer, 1993: 118)

In the case of mathematics, the shift from the value of meaning towards the value of utilisation is often paralleled by a process in which mathematical knowledge materialises into technologies (social technologies included), thus becoming implicit knowledge. Demathematisation, in this context, denotes the phenomenon of mathematics having lost its value of meaning for the majority of people. For most individuals mathematics is of no help for existential questions. At the same time the value of utilisation is obscured – mathematics has become invisible: not even for using all the mathematics-based technology it is necessary to be mathematically competent. More than ever, however, people make use of "crystallized" mathematical knowledge, but instrumental mathematical competence is not required.

Bishop, in this volume, reports on a study of mathematics and science teachers' values based on a model comprising of six sets of value clusters that are structured as three complementary pairs. The dimension of "control and progress" describes how individuals relate to their mathematical and scientific knowledge. These values resemble the ways in which individuals interpret the value of utilisation.

It can be argued, as Fischer (1993) does, that the value of utilisation, obscured in and through technology, is well connected to the value of meaning. The mathematical construction of many parts of reality constrains the ways in which people give meaning to their life. Following this line of argument, mathematics still is meaningful for people, but not in an explicit sense. Mathematics, in Fischer's (1993) terms, is a *means* of control and manipulation of the natural and the artificial environment and a *system* of concepts and rules, *embodied* in our thinking and doing. While mathematics, as a means, is conceptually similar to the value of utilisation, the system aspect goes beyond the value of meaning. The system aspect has explanatory power with respect to the relation of mathematisation and demathematisation. It covers the fact that mathematics has transformed into a system that:

> runs from everyday quantifications to elaborated patters of natural phenomena to complex mechanisms of the modern economy. … [This is a system] we have to obey and which is inseparably connected with our social organization. (Fischer, 1993: 114).

The value of meaning can be regarded as the conscious part of mathematics as a system. The unconscious part, however, is at the core of the relation of mathematisation and demathematisation. Mathematics has the potential of contributing to more consciousness of society, if it were not used dogmatically for legitimating the given, but would instead be interpreted as a means for provoking decisions, which might lead to change, as Fischer argues in this volume.

An important but often neglected aspect is illuminated by Davis in this volume. He presents an analysis of the public image of mathematics conveyed by the media. It shows that the value of utilisation is obscured and the public image consists of an imperceptive view of the fact that mathematisation "is changing what it means to be human"; mathematics is mystified by concentration on the sensational. As Ahmed, in this volume, argues, students in school usually gain little experience of the power of mathematics, not even of the "sensational". They are likely to develop a narrow, techniques and convention oriented view of the subject. Malcolm, in this volume, makes a related observation of the way science is taught at school. "School science, traditionally, has taken the Enlightenment position, promoting and demonstrating reason, empiricism, objectivity and mathematical logic as the key characteristics of science." Restricting school science to these values is one of the reasons for exclusion of distinct groups of students and thus counteracting the vision of 'Science for All'.

Mathematisation/ demathematisation and power

Skovsmose (1998) regards mathematics as an essential instrument for exercising technological power. He sees an increase in the range of applications of mathematics linked to modern information technology. Mathematics has not only become an integrated part of technological planning and decision making but also an invisible part of social structuration, encapsulated in political arguments, technologies and administrative routines. Citizenship presupposes the excavation of "frozen" mathematics.

Following this line of argument, demathematisation excludes citizenship, and development of appropriate excavation tools becomes a central issue. Skovsmose introduces positions of social groups who are involved in or affected by mathematics in action in different ways. The "constructors" are those who "develop and maintain the apparatus of reason" (Skovsmose, 2006: 140). In constructing mathematics based technology, this group exercises power over "operators" and "consumers" of this technology. Vithal, in this volume, discusses issues of ethics and political responsibility and of social and cultural awareness of those who occupy positions of power. This affects the mathematics and science curricula of graduates as well as the opportunities of researchers. However, the so called "math wars" in the United States show, as Kilpatrick reports in this volume, that mathematicians oppose the standards developed by the Nation Council of Teachers of Mathematics – these standards being an attempt of providing a vision of mathematical literacy for today's world. Their worries are, for example, directed towards "schoolchildren wasting their time making histograms of data they have gathered when they could be learning arithmetic. They see statistics in school mathematics as involving little serious work."

Whereas the constructors are involved in developing mathematical technology, the operators are those who work in jobs, in which they have to make decisions on the input and then decisions based on the output of this technology. These job situations can be called "rich in implicit mathematics" (Skovsmose, 2006: 142). Operators are not only prepared for their tasks in terms of the content of their mathematical training, but also accustomed to the "habit" of following rules by the hidden curriculum of school mathematics. Skovsmose (in press) calls those who are listening to a range of offers, statements and reports containing figures and numbers, slightly ironically, "consumers" of mathematics. They could "vote, receive services, fulfil obligations, be citizens". Consumers are confronted with justifications of decisions based on complex models. While the consumers can be regarded as "targets" of mathematisations, there is still another group of people Skovsmose calls the " 'disposable' ", those who are marginalised and who "do not define themselves with reference towards what they miss in a relationship with the globalised world" (Skovsmose, in press).

There is a threat to democracy because of a widening gap of mathematical knowledge between constructors and consumers. The constructors not only provide the technical knowledge for developing solutions but also have the power to define the problems and to initiate new questions. The forming of opinions and political decisions become more and more dependant on their expertise.

Skovsmose (1994) sees as one of the essential problems of democracy in a highly technological society, the development of a critical competence, which can match the actual social and technological development. If the interpretation of democracy is not restricted to formal procedures of electing a body of representatives, but also includes participation and elements of direct democracy, the status of the constructors has to be scrutinised. Decisions made on the ground of mathematical models may be inaccessible to demathematised consumers.

> Citizenship does not only imply being ready to live in and to face the "output" from authorities. It also means providing an "input" to authority, a "talking back" to authority. (Skovsmose, 1998: 198-99).

As to mathematisations, 'talking back' presupposes a wider horizon of interpretations and pre-understandings of mathematical knowledge than passive consumption of offers, statements and reports:

> The fundamental thesis relating technological and reflective knowledge is that technological knowledge itself is insufficient for predicting and analysing the results and consequences of its own production; reflections building upon different competencies are needed. The competence in constructing a car is not adequate for the evaluation of the social consequences of car production. (Skovsmose, 1994: 99).

Skovsmose (1998) identifies three groups of questions related to reflective knowledge, which focus (i) on the relationship between mathematics and an extra-mathematical reality, (ii) on mathematical concepts and algorithms, and (iii) on the social context of modelling and its implications in terms of power.

The distinction between technological and reflective knowledge, which also resembles the distinction between operators and critical consumers (in opposition to demathematised consumers), is fruitful, but still has to be further elaborated with respect to its consequences for a conceptualisation of content and forms of mathematics education. Especially with respect to those groups of people who are deprived of any kind of formal education (the 'disposable'), the tension between functional and critical education seems to be exacerbated.

Skovsmose, in this volume, asks in which ways mathematics, which might be conceived as important for their future lives by some students, but not necessarily by those who are already marginalised, could offer new opportunities. Mathematics education should not only try to relate to the students' background, but to their foreground, that is the opportunities provided by the social, political and cultural situation as conceived by a student. Also, the discourse of learning obstacles is linked to students' (social and cultural) background rather than to their foreground, which shows the political nature of the concept.

REMARK

This introductory chapter has reviewed a selection of literature, which explicitly refers to mathematisation/ demathematisation and discusses this phenomenon as a social process. An attempt has been made to draw out issues related to the formatting power of mathematics and its role as implicit knowledge, resulting in a process of demathematisation – a concept that, after having received considerable attention, is now threatened to be eclipsed by the proliferation of a discussion of school mathematics, which shows a tendency of cutting off its own philosophical and political roots.

NOTES

[1]   This example has been adapted from the WebPage of the King's College of the University of London: http://www.kcl.ac.uk/phpnews/wmview.php?ArtID=661

[2]   Another example of technology Keitel repeatedly uses in order to illustrate how the development and transfer of technology results in demathematisation is the economic instrument of double-entry bookkeeping (e.g., Keitel, 1989; Keitel, Kotzmann, & Skovsmose, 1993; Damerow, Keitel, Elwitz, & Zimmer 1974):

> Because of its enormous diffusion the double-entry model penetrates practically all fields of social practice related to money. And like crystallisation in a liquid, which starting from one point expands over the whole surface, the calculation model sets going systematization and formalization all over the area where it is applied. In industrial enterprises where for a long time production followed its own traditional patterns, "scientific management" and "system analysis" ultimately led to the restructuring of all production processes in the most minute detail towards the goal of systematization and standardization in order to bring them under the control of the calculation model and its prescriptions. (Keitel, 1989: 11-12).

REFERENCES

Adler, J. (2001). *Teaching mathematics in multilingual classrooms.* Dordrecht: Kluwer.

Bernstein, B. (1996). *Pedagogy, symbolic control and identity: Theory, research, critique.* London: Taylor & Francis.

Bernstein, B. (1999). Vertical and horizontal discourse: An essay. *British Journal of Sociology of Education, 20,* 157-173.

Chevallard, Y. (1989). Implicit mathematics: Its impact on societal needs and demands. In J. Malone, H. Burkhardt & C. Keitel (Eds.), *The mathematics curriculum: Towards the year 2000: Content, technology, teachers, dynamics.* Perth : Curtin University of Technology, 49-57.

Coben, D. (Ed.) (2003). *Adult numeracy: Review of research and related literature.* London: National Research and Development Centre for adult literacy and numeracy.

Cooper, B., & Dunne, M. (2000). *Assessing children's mathematical knowledge: Social class, sex and problem-solving.* Buckingham: Open University Press.

Damerow, P., Keitel, C., Elwitz, U., & Zimmer, J. (1974). *Elementarmathematik: Lernen für die Praxis? Ein exemplarischer Versuch zur Bestimmung fachüberschreitender Curriculumziele.* Stuttgart: Klett.

Davis, P. J. (1989). Applied mathematics as social contract. In C. Keitel, P. Damerow, A. J. Bishop & P. Gerdes (Eds.), *Mathematics, education, and society.* Paris: UNESCO, 24-28.

Davis, P. J., & Hersh, R. (1986). *Descartes' dream: The world according to mathematics.* San Diego: Harcourt Brace Jovanovich.

de Lange, J. (1996). Real problems with real world mathematics. In C. Alsina, J. M. Alvarez, M. Niss, A. Pérez, L. Rico & A. Sfard (Eds.), *Proceedings of the 8th International Congress on Mathematical Education.* Sevilla: S.A.E.M. Thales, 83-110.

de Lange, J., Keitel, C., Huntley, I., & Niss, M. (Eds.) (1993). *Innovation in maths education by modelling and applications.* Chichester: Ellis Horwood.

Dowling, P. (1996). A sociological analysis of school mathematics texts. *Educational Studies in Mathematics, 31,* 389-415.

Dowling, P. (1998). *The sociology of mathematics education. Mathematical myths/pedagogic texts.* London: RoutledgeFalmer.

Fischer, R. (1993). Mathematics as a means and as a system. In S. Restivo, J. P. van Bendegem & R. Fischer (Eds.), *Math worlds: Philosophical and social studies of mathematics and mathematics education*. New York: SUNY Press, 113-133.

FitzSimons, G. E. (2002a). Critical mathematical literacy for adult and vocational students. In L. Bazzini & C. Inchley (Eds.), *Mathematical literacy in the digital era*. Milano: Ghisetti e Corvi, 137-141.

FitzSimons, G. E. (2002b). What counts as mathematics? Technologies of power in adult and vocational education. Dordrecht: Kluwer.

Freudenthal, H. (1983). *Didactical phenomenology of mathematical structures*. Dordrecht: D. Reidel.

Gates, P., & Vistro-Yu, C. P. (2003). Is mathematics for all? In A. J. Bishop, M. A. Clements, C. Keitel, J. Kilpatrick & F. K. S. Leung (Eds.), *Second international handbook of mathematics education*. Dordrecht: Kluwer, 31-74.

Gellert, U., Jablonka, E., & Keitel, C. (2001). Mathematical literacy and common sense in mathematics education. In B. Atweh, H. Forgasz & B. Nebres (Eds.), *Sociocultural research on mathematics education: An international perspective*. Mahwah: Erlbaum, 57-73.

Gravemeijer, K. (1994). *Developing realistic mathematics education*. Utrecht: CD- β Press.

Harris, P. (1991). *Mathematics in a cultural context: Aboriginal perspectives on space, time and money*. Geelong: Deakin University Press.

Houston, S. K., Blum, W., Huntley, I., & Neill, N. (Eds.) (1997). *Teaching and learning mathematical modelling: Innovation, investigation and applications*. Chichester: Albion.

Jablonka, E. (1997). What makes a model effective and useful (or not)? In S. K. Houston, W. Blum, I. Huntley & N. Neill (Eds.), *Teaching and learning mathematical modelling: Innovation, investigation and applications*. Chichester: Albion, 39-50.

Jablonka, E. (2003). Mathematical literacy. In: A. J. Bishop, M. A. Clements, C. Keitel, J. Kilpatrick & F. K. S. Leung (Eds.), *Second international handbook of mathematics education*. Dordrecht: Kluwer, 75-102.

Keitel, C. (1989). Mathematics education and technology. *For the Learning of Mathematics, 9*(1), 103-120.

Keitel, C., Kotzmann, E., & Skovsmose, O. (1993). Beyond the tunnel vision: Analysing the relationship between mathematics, society and technology. In C. Keitel & K. Ruthven (Eds.), *Learning from computers: Mathematics education and technology*. Berlin: Springer, 243-279.

Marx, K. (1961 / 1867). *Capital*, vol. 1. Moscow: Foreign Languages Publishing House.

Matos, J. F., Blum, W., Houston, S. K., & Carreira, S. P. (Eds.) (2001). *Modelling and mathematics education*. Chichester: Horwood.

Mumford, L. (1967). *The myth of the machine, vol. 1: Technics and human development*. New York: Harcourt Brace Jovanovich.

Skovsmose, O. (1994). *Towards a philosophy of critical mathematics education*. Dordrecht: Kluwer.

Skovsmose, O. (1998). Linking mathematics education and democracy: Citizenship, mathematical archaeology, mathemacy and deliberative education. *Zentralblatt für Didaktik der Mathematik, 98*(6), 195-203.

Skovsmose, O. (2006). *Travelling through education: Uncertainty, mathematics, responsibility*. Rotterdem: Sense.

Skovsmose, O. (in press). Mathematical literacy and globalisation. In B. Atweh, M. Borba, A. Calabrese Barton, D. Clarke, N. Gough, C. Keitel, C. P. Vistro-Yu & R. Vithal (Eds.), *Internationalisation and globalisation in mathematics and science education*. Berlin: Springer.

Sohn-Rethel, A. (1978). *Intellectual and manual labor: A critique of epistemology*. London: Macmillan.

Streefland, L. (Ed.) (1991). *Realistic mathematics education in primary school*. Utrecht: CD-β Press.

Taylor, F. W. (1947). *Scientific management*. New York: Harpers and Brothers.

Tenbruck, F. H. (1975). Der Fortschritt der Wissenschaften als Trivialisierungsprozeß. *Kölner Zeitschrift für Soziologie und Sozialpsychologie/Sonderheft 18*, 14-47.

Treffers, A. (1987). *Three dimensions. A model of goal and theory description in mathematics instruction – the Wiskobas project*. Dordrecht: D. Reidel.

17

Treffers, A. (1991). Didactical background of a mathematics programm for primary education. In L. Streefland (Ed.), *Realistic mathematics education in primary school.* Utrecht: CD-β Press, 21-56.

Treffers, A., & Goffree, F. (1985). Rational analysis of Realistic Mathematics Education. In L. Streefland (Ed.) *Proceedings of PME-9.* Utrecht: OW&OC, 97-123.

Vithal, R., & Skovsmose, O. (1997). The end of innocence: A critique of 'ethnomathematics'? *Educational Studies in Mathematics, 34*(2), 131-157.

AFFILIATIONS

Eva Jablonka
Fachbereich Erziehungswissenschaft und Psychologie,
Freie Universität Berlin

Uwe Gellert
Fachbereich Erziehungswissenschaft,
Universität Hamburg

PETER DAMEROW

THE MATERIAL CULTURE OF CALCULATION

A Theoretical Framework for a Historical Epistemology
of the Concept of Number

INTRODUCTION

Reflections on numbers and their properties led already in antiquity to the belief that propositions concerning numbers have a special status, since their truth is dependent neither on empirical experience nor on historical circumstances. In a historical tradition extending from the Pythagorean through the Platonic tradition of Antiquity, Late Antiquity and the Middle Ages, further through the rationalism and the critical idealism of Kantian and neo-Kantian philosophy to the logical positivism and constructivism of the present, this belief has been considered proof that there are objects of which we can gain knowledge *a priori*. Like a recurring leitmotif, the conviction that numbers are by nature ahistorical and universal is woven through the history of philosophy. A variety of reasons have been proposed to explain this puzzling phenomenon.[3] The historian, on the other hand, is confronted with the fact that numerical techniques and arithmetic insights have a history that is, at least on its surface, in no way different from other achievements of our culture.[4] In view of the variety of historically documented arithmetical techniques, it is scarcely possible to dismiss the assumption that the concept of number – in the same way as most structures of human cognition – is subject to historical development, which in the course of history exposes it to substantial change.

If we ask which of the two alternatives is correct or if, possibly, the alternative itself has to be called into question, we must remember that this is a matter not considered solved in either of the disciplines dealing with the number concept, but rather, on the contrary, one which has for some time been the object of fundamental and ongoing controversy. This is particularly true in the case of psychology.

In the initial phase of modern psychology under the influence of neo-Kantianism, numbers and similar mathematically determined objects were primarily regarded as results of thought processes found in all humans alike. Only with the reaction of Gestalt psychology (Wertheimer, 1925) to the challenge of universal concepts of number by cultural anthropology (Lévy-Bruhl, 1926) did the question of the nature of numbers find its way into psychology. In particular the empirical evidence, provided by Piaget, that the concept of number is not already imprinted in a child at birth, but is rather formed during the development of the

U. Gellert, E. Jablonka (eds.), Mathematisation – Demathematisation: Social, Philosophical and Educational Ramifications, 19–56. © 2007 Sense Publishers. All rights reserved.

child in a number of developmental stages (Piaget, 1952), contributed to undermining the belief in the innate nature of the concept of number. Piaget himself, however, still interpreted his results entirely in the spirit of neo-Kantianism. In his theory the development of the concept of number in ontogenesis rests on experience; however, the result of the development, according to this theory, is determined epigenetically – similar to many other biologically determined characteristics of humans. Numbers are thus considered as a cross-culturally universal cognitive construction that only appears at the end of the development of the individual (Piaget, 1950, 1971). Ethnographic research, conducted in the tradition of the developmental psychology that reflects Piaget's work and methods, has predominantly used this premise as its point of departure, and has accordingly arrived for the most part at the conclusion that the speed of the development of logico-mathematical thought varies markedly under diverse sets of cultural circumstances, not so however the structure of the logico-mathematically structured concepts themselves (Piaget & Garcia, 1989; Bruner, Olver, & Greenfield, 1966; Dasen & Heron, 1981; Dasen & Ribaupierre, 1988; Hallpike, 1979).[5] Piaget offered a number of arguments suggesting that cognition in primitive cultures that have no developed arithmetical activities is comparable to the pre-logical stage of the ontogenetic development of a child, in which cognition cannot yet make use of the mental operations that are characteristic of the subsequent concrete operational stage of ontogenesis (Piaget 1950, vol. 2: 72-82).[6] He distinguished two fundamentally different phases of development for each logico-mathematical concept: an initial phase in which the historical development passes through universal stages that are ontogenetically identifiable, and a second one in which the development is no longer subject to universal laws, but rather to a historical logic of development constituted by reflective abstractions (Damerow, 1995: 299-370).

Contrary to such universalistic interpretations of the concept of number, the examination of the particular mental processes connected with arithmetic operations led to theoretical approaches in which the emergence of numbers appears as the result of manifold learning processes (Gelman & Gallistel, 1978; Brainerd, 1979; Fuson & Hall, 1983; Smith, Greeno, & Vitolo, 1989; Gallistel & Rochel, 1992). Modern cognitive science has increasingly supported this view, recently by providing evidence that many arithmetic accomplishments can be attributed to the construction of relatively simple "mental models".[7] Further alternatives came into the discussion through the work of psycholinguists and their interpretation of the concept of number as a linguistic phenomenon, without, however, bringing the question closer to resolution. Under the influence of Chomsky's theory, numbers have been ascribed to a biologically determined syntax of language (Hurford, 1975, 1985). Psycholinguistic investigations of the representation of logico-mathematical structures in language on the other hand suggest that we might understand such structures and the objects constituted by them better as culturally relativistic constructions.[8]

Such contradictions between different conceptions of number obviously cannot be solved within the limited point of view of a single discipline, since neither a study of the cognitive functions of the concept of number excluding the question of its historical changes, nor a study of the historical development of arithmetical techniques leaving out of consideration the cognitive functions of those techniques, do justice to the unsolved problems that are revealed in these controversies. In what sense does the concept of number represent a universal? In what respect is it subject to historical changes? What implications result for the relation of the ontogenetic development of the concept of number to the historical changes of numerical techniques and arithmetical insights? These questions can only be answered by a historical epistemology of arithmetical thought that is compatible with psychological theories as well as with the results of historical research (Damerow, 1998). This view of the problems determines the theoretical program to be outlined in the following.

PRINCIPLES OF A HISTORICAL EPISTEMOLOGY OF
LOGICO-MATHEMATICAL THOUGHT

On the nature of historical developmental processes of cognition

Since the historical development of cognition is realized through the cognitive activity of individuals, the description of cognitive abilities in the study of their historical development cannot, in principle, differ from that in the study of their individual development. Problematical, however, is the transfer of psychological concepts to historical development in the case of the developmental processes, insofar as individual development of cognitive knowledge representation structures is a process fundamentally different from their historical development.

The *individual* development of cognition is a process in the mind of the individual person. It starts with the awakening of intelligence in childhood and ends with the death of the person. The *historical* development of cognition, however, is a collective process spanning populations and generations, based on the interaction of individuals whose minds are basically independent of each other. The process of transmitting cognitive knowledge representation structures from one generation to the next takes place in a network of individual paths of tradition, leading from the individuals of one generation to the individuals of the next. These transmissions are realized in symbolic, and in part also in direct, interactions. There are no obvious reasons to assume that the network of those paths of tradition might show analogies to the individual development of cognition. The historical development of knowledge representation structures is by its nature a phenomenon that has to be interpreted socio-historically and not psychologically.

Nonetheless, the historical development of such structures is based on interactive processes, founded on very particular conditions that can be described psychologically. Not every individual process of achieving and structuring knowledge influences the historical development of cognition. Results of individual cognitive processes that are not transferable mental models which can be adopted by others in socialization processes, are obviously largely irrelevant to historical developments. Likewise, the results of universal ontogenetic processes of

21

development naturally cannot exhibit historical changes that might lead to coherent lines of development in the paths of tradition constituted by interactions. The network of those paths of tradition of cognitive abilities can apparently only then be subject to coherent processes of development when, in social interaction, results of individual cognitive processes are systematically reproduced and extended by consecutive generations.[9] The reproduction of culture-specific mental models is therefore the most important psychologically describable condition for historical processes of cognition.

Basic assumptions concerning the development of logico-mathematical thought

The theory of the historical development of arithmetical thought proposed here is essentially based on two assumptions: *Firstly*, it is assumed, following Piaget's genetic epistemology, that logico-mathematical concepts are abstracted not directly from the objects of cognition, but from the coordination of the actions that they are applied to and by which they are somehow transformed. According to this assumption the emergence of mental operations of logico-mathematical thought is based on the internalisation of systems of real actions. The internalised actions are the starting-point for meta-cognitive constructions, through which they become elements of systems of reversible mental transformations which, following Piaget's terminology, we will call here 'operations'. Meta-cognitive constructs such as the concept of number that are generated by reflective abstractions can thus be understood as internally represented invariables of mental operations which reflect actions on real objects. This explains the puzzling *a priori* nature of constructions such as the number concept. The experience of objects appears to be preformed by logico-mathematical structures. These structures, although they are subject to processes of development that have their origin lastly in the experience of objects, can therefore no longer be changed by those experiences.

Secondly, differing from Piaget's theory, it is presumed that the basic structures of logico-mathematical thought are not determined epigenetically, but are developed by the individual growing up in confrontation with culture-specific challenges and constraints under which the systems of action have to be internalised. The challenges are embodied in the material means of goal-oriented or symbolic actions that are shared external representations of logico-mathematical structures of mental models. These mental models, upon which logico-mathematical competence is based, are thus construed in ontogenesis conditioned by the processes of socialization and by the co-construction of such models by means of interaction and communication. These co-constructions on the micro level of social interaction make it possible that mental models are transferred from one individual to the next and so, on the macro level of social development, are transmitted as intersubjectively shared schemata of interpretation. This process gains historical continuity through collective external representations which embody both mental models and levels of reflection, and thus levels of abstraction.

These two assumptions specify preconditions for a theory of the historical development of logico-mathematical thought. By combining both assumptions, a twofold result is achieved. On the one hand, certain objects of psychological theories receive a historical interpretation; on the other, historical stages of the development of thought can be characterized psychologically. The stages of the historical development of logico-mathematical thought can be interpreted as subsequent meta-cognitive levels that are connected with each other by reflective abstractions.

On the definition of historical stages of development

The application of psychological concepts to historical processes raises, however, a number of fundamental problems. One of these problems concerns the definition of historical stages of the development of cognition in general. Since psychological definitions of abilities by their nature do not refer to collective subjects, psychologically defined abilities cannot readily characterize historical stages of development. They can only be attributed to an individual person, to members of a group, or to all members of the society in a particular historical situation, not however to the society as a whole. A definition of historical stages of development using psychological concepts seems scarcely possible without determining, with a certain arbitrariness, on which of these different distributions of competence the definition of a historical stage as a criterion of its realization is to be based.

The second of the basic assumptions formulated above offers, however, a solution for this problem (Damerow, 2000). If the historical development of cognitive knowledge representation structures is essentially based on their intersubjective communication and historical transmission by means of external representations of mental models, then the social distribution of the competence is of only secondary importance for this development. It then represents only a framing condition, determining above all the speed of development and the chances of realization for the cognitive potentials embodied in the representations. The historical stages of development, on the other hand, have to be defined primarily on the basis of analyses of such representations and their possible functions in the individual development of cognition.

The theory of the historical development of logico-mathematical thought outlined here is therefore not meant to explain primarily the outstanding achievements of individuals nor the social distribution of abilities, but rather the historically changing potentials of development of the individual subject under the conditions prevailing at the time. In particular, the level of development of arithmetical thought in the various cultural epochs is not being defined by the actual results of arithmetical thought, but by the arithmetical means and external representations of mental models that were available for the ontogenetic development of arithmetic abilities, so that these could in principle evolve.

The resulting theory of the historical development of knowledge representation structures can be applied as well to the analysis of the construction of new representations which result from outstanding individual achievement as also to the application of such a representation by a specifically trained group, or even to the general use of such a representation in a society whose educational system communicates such use. Its application to the development of cognition can explain various aspects of historical stages defined by representations with certain cognitive characteristics. The theory allows to infer that at these stages certain individual accomplishments become possible, certain professional qualifications become reasonable and certain goals of education become generally understandable.

The transition from one stage to the next

In the historical development of cognition, the transitions from one stage of the development of certain knowledge representation structures to the next one can occur in two fundamentally different forms, namely either by cultural exchange or by culture-immanent processes of construction.

The diversity of cultures coexisting and interacting with each other usually results in the adoption of representations that have shown themselves to be effective tools of cognition in another culture. Such processes of transmission have to be studied in order to find out which of the cognitive knowledge representation structures found in many, or even in all cultures are biologically inherent in human nature[10] and which, on the other hand, are a result of a transfer of representations of these structures to many or all existing cultures.[11]

Determinative for defining historical stages of development in the processes of the historical genesis of knowledge representation structures is, however, the development of mental models by culture-immanent constructions. This form of development is based first on individual cognitive achievements that lead to the modification of existing representations and to the construction of new ones. These representations become part of a culture by being embedded in existing paths of tradition, so that they can be integrated into the process of reproduction in this culture.

In both cases, interactive co-constructions and transmissions of mental models from one individual to the next constitutes a necessary precondition for the emergence of a new stage of development. The individual, creative achievement, which historians of science tend to credit with a crucial role in the rise of new forms of thought, is only a peripheral condition of this development, which is rather determined by existing representations of knowledge and by contingent historical circumstances. Any historical situation defines a space of potential cognitive achievements which enables individual creative achievements and, at the same time, imposes narrow limits on them. The question of how the meaning of historically determined representations of knowledge can be reconstructed by an individual reflecting on his activities in a given cultural setting becomes a theoretical key question for the understanding of the historical development of cognition.

Two kinds of representations of mental models will be distinguished that are fundamentally different with regard to the level of reflection crucial for their meaning. The former will be called first-order external representations, the latter second-, or more generally, higher-order external representations. The difference consists, briefly stated, in the fact that first-order external representations stand for real objects and actions, higher-order external representations, however, for ideas and mental activities. How can such a differentiation be theoretically specified?

First-order external representations

Definition: First-order external representations of mental models (or briefly: first-order representations) are material representations of real objects by symbols or by models composed of symbols and rules of transformation, with which essentially the same actions can be performed as with the real objects themselves.[12]

Some simple examples may make this definition plausible. The identification of a concrete object with a name, a word or a sign is a first-order representation of the object. Identifying such objects and their properties as permanent and grouping them with other objects can be performed with such representations in the same way as with the objects themselves. Another example is provided by counters and similar symbolic counting aids which represent real objects in one-to-one correspondences. Words used for counting and similar symbols such as body numbers that can be arranged in temporal and spatial succession are first-order representations of ordinal structures such as sizes or intensities of properties as, for instance, the shades of a colour range. They can serve to perform comparisons on the level of symbolic representation in order to compare sizes of the represented objects or to determine maxima and minima in a given real setting. Constructions with compass, ruler and similar graphic instruments are first-order representations of geometrical configurations in the Euclidean plane that can be used for identifying locations of objects, spatial relations between them, and changes of such relations when these objects are moved. The cutting and pasting of areas performed in a suitable medium of geometrical representation is a first-order representation of the additive structure of real areas such as, for instance, cultivated fields that are assigned to owners, used to grow specific plants, or selected for the allocation of the water from an irrigation system.

The usefulness of first-order representations is based on the fact that actions can as a rule be performed much more easily with the symbols of the representation than with the real objects they represent, since they are not, to the same degree, subject to accidental restrictions characteristic of real situations. Obviously, real actions represented by such symbolic actions cannot be substituted by them. Joining two groups of tokens representing two flocks of sheep, for instance, cannot substitute joining the two flocks themselves. The representation of sheep by tokens rather serves as a tool for performing symbolic actions anticipating the results of real actions in order to plan and control them.

First-order representations, however, are not only significant for the execution of operations of existing mental models but also contribute to their construction by the internalisation of systems of actions. Unlike in the case of their simple application to directly control real actions, symbolic actions in their role of supporting the construction of mental models can completely substitute the real actions, because they share essential physical qualities with the objects and actions for which they stand. Symbolic actions in the system of rules of a first-order representation thus initiate the construction of the same mental model as actions with the real objects they represent. This quality of a system of actions will be designated in the following by the expression constructive. Symbolic actions that are performed with first-order representations are in this sense constructive with respect to the mental model that controls the real actions they represent.

First-order representations are abstract in the sense that the same symbols are used in diverse contexts. This leads to a differentiation in the meaning of symbols characteristic for this kind of representation. A symbol in a first-order representation embodies a concrete object which changes from application to application and, at the same time, an abstract object implicitly defined by the operations of the constructed mental model that remains the same in all applications.

Second- (or higher-) order external representations

Definition: Second- (or higher-) order external representations (or simply second-order representations) are material representations of mental models. They consist of symbols or models composed of symbols and rules of transformation which correspond to the operations of the abstract mental model that controls the actions performed with the real objects.

The adequate application of a second- (or higher-) order representation requires that it gets related to the real objects and actions to be represented by its symbols and transformations. This happens by assimilating these objects and actions to the mental model that gives the external representation its meaning.

Again, some simple examples may make this definition plausible. Words that are conventionally assigned to numbers (one, two, three, ...) and non-constructive numerals (1, 2, 3, ...) become second-order representations of abstract numbers once the developmental stage has been reached at which this concept occurs. Their application to real objects requires thus the understanding of the concept of number. The use of the word "number" itself and of any general terminology related to the attributes of numbers,[13] the use of variables as universal numbers, and even abstract calculations with numerals independent of concrete applications are also examples of the use of higher-order representations of the number concept. Their applicability is based on the reflective manipulation of numbers and their representations, for example on the correct use of substitution rules for variables.[14] Furthermore, certain theorems of Euclid's *Elements* such as the Pythagorean theorem are second-order representations of the metric structure of the Euclidean plane. Albeit propositions on geometric figures, they are independent of these figures insofar as the objects they relate to are no longer these concrete figures but rather "virtual" mathematical objects that are implicitly defined by axioms and definitions within a framework of deductive representation.

Second-order representations represent real objects and actions only indirectly. They are not constructive with regard to the mental model they represent since their adequate application to real objects and actions presupposes that this model

26

exists already. They are, however, constructive with regard to meta-cognitive mental models that may be constructed as higher-order representations by reflective abstraction as far as they are directly related to material actions performed on the real objects and actions (signs and sign transformations) of first- (or lower-) order representations which are the basis of the mental models they represent.

This shows the precise connection between second- (and higher-) order representations with reflective abstractions which, according to the first assumption formulated above, create logico-mathematical concepts. Since the material symbols and symbolic actions of second- (and higher-) order representations can themselves become objects of cognition, they initiate precisely that kind of concept formation for which Piaget introduced the term 'reflective abstraction'.

Second-order representations result in a similar way as first-order representations in a differentiation of the meanings of symbols. While first-order representations initiate the differentiation between the concrete objects of real actions and abstract objects of mental models, second-order representations initiate the differentiation between such abstract objects and those objects which are implicitly defined by the symbols and symbol transformations of this second-order representation, which are abstract objects of a meta-cognitive nature.[15]

The historical change of cognitive functions of external representations

As a rule, representations change their function in the process of historical development as well as in individual cognitive development. In particular, higher-order representations develop from first-order representations.

All systems of counting, for example, were originally first-order representations. They mainly represented ordinal structures, as a rule by the temporal succession of a conventionally determined counting sequence. They represented cardinal structures only insofar as by creating in the process of counting a one-to-one correspondence of real objects with words or body parts they could also be used as a first-order representation to identify cardinal quantities. With the development of the number concept, more and more abstract numerical properties became incorporated into their meaning. They thus developed into second-order representations of abstract numerical entities and arithmetic rules. In particular, they now also represented structures like multiplication which have no immediate basis in the symbolic action of counting. When the abstract concept of number had finally developed, the words used in counting developed into the abstract infinite counting sequence, the meaning of which incorporated, step by step, all deduced abstract properties of numbers such as, for instance, the infinity of the sequence of prime numbers. The counting sequence thus became a higher-order representation of the abstract concept of natural numbers.

The change of the function of representations was in a similar way also characteristic of the development of geometry. The prehistory of deductive geometry was shaped by the use of drawings and later also of true-to-scale constructions as first-order representations of relationships in empirical space.

Constructions were still playing an essential, though different role in Euclid's *Elements*. The ancient version of Euclid's geometry comprised not only theorems with the proofs of their truth, but also so-called problems consisting of constructions and proofs that the constructed figures possess the required properties. This duality of constructions and proofs in Euclid's *Elements* indicates that figures still served here as first-order representations complementing the deductive second-order representation in written language. In later editions and revisions of Euclidean geometry, the problems degenerated into helpful but basically dispensable exercises. The required information was now derived from mental images of the real figures which thus developed into second-order representations of the geometrical relations represented by such generalized mental images. The apparent independence of the mental images from the real figures seemed to show the a priori nature of geometrical knowledge. The construction of real figures developed into a meta-cognitive tool of judgments about the epistemological function of Euclidean proofs. This finally led to the construction of non-Euclidean geometries and the modern formalism of proof techniques, thus disproving the alleged a priori nature of geometrical knowledge. Meta-mathematical proofs of the consistency of non-Euclidean concepts now used geometrical figures as Euclidean models of abstract structures, i.e., as higher-order geometrical representations.[16]

The historical transmission of the meanings of external representations

Two basic assumptions have been introduced at the beginning of this paper which are constitutive for the theory presented here. First, it is assumed that logico-mathematical concepts are abstracted invariants of transformations, transformations which are realized by actions, and second, that those abstractions are historically transmitted by collective external representations. These processes come about in culture-specific, historically changing symbolic scenarios. The key question to be answered is the question of how the members of a community can reconstruct and further develop the meaning of such historically transmitted external representations.

Since the mere reconstruction of the meaning of first-order representations does not yet require specific cognitive structures of logico-mathematical thought, they play a special role: first-order representations are the starting-point of abstraction and thus to a certain extent determine the structure of the semantics of second- (and higher-) order representations generated by reflective abstraction. In the same way that first-order representations of real objects and actions initiate the construction of new mental models, they also make possible the reconstruction of those mental models which already exist historically and which give a specific meaning to culture-specific first-order representations.

Second- (and higher-) order representations, that is, external representations of such mental models, require, however, that these mental models exist already. They must have been constructed or reconstructed already by means of first-order or lower-order representations respectively. As soon as this condition is established second- (and higher-) order representations have the same function in the reconstruction of the reflected meaning of mental transformations as first-order representations have for the meaning of concrete logico-mathematical activities themselves.

The historical development of logico-mathematical thought is thus based, according to this theory, on two psychologically explicable processes, namely on the individual construction and on the ontogenetic reconstruction of the meanings of representations. The differentiation of the functions of first-order and higher-order external representations provides a powerful theoretical tool for the explanation of historical as well as individual processes in the development of logico-mathematical thought.

Three different processes and their combined effects have to be examined, the ontogenetic emergence of seemingly universal cognitive structures which are independent of culture-specific representations, the reconstruction of the meaning of first-order representations, and the reconstruction of second- (and higher-) order representations.

First, there are systems of actions which develop on the basis of the biological potentials of humans and can consequently be found in all cultures. Those systems contain, for example, actions such as the grasping of objects, the arranging of objects, the movement of objects or of the own body in space, etc. These most general human activities correspond to the universals of logico-mathematical thought. Such universals are the basis of supercultural mental models that are not subject to historical changes.[17]

Second, there are very general systems of actions that are not determined by the human biological potentials but include the use of simple tools which are used in all or almost all cultures. They are therefore subject to historical changes, and are not found in all cultures in the same way. Once these systems exist, they exhibit in their simplicity little if any fundamental variation and are therefore largely transmitted by the same kind of first-order representations.[18] To these systems of actions belong, for example, the techniques of counting which are manifest in almost all cultures.

Third, there are countless culture-specific systems of actions and higher-order representations that are of importance for the reconstruction of the historical development of logico-mathematical thought since without them the complex mental models of culture-specific forms of abstract thinking that they embody cannot evolve in ontogenesis.[19]

How can it be explained that the mental models of logico-mathematical thought constructed and transmitted this way by means of external representations seem to be logically determined, resulting from knowledge *a priori*? These mental models apparently are no longer connected to the empirical knowledge that constituted the starting-point for their construction. If the theory of the function of higher-order representations submitted here is correct, there is a simple explanation for this characteristic. The independence of implicitly defined objects may be interpreted as the immediate consequence of the relative independence of higher-order representations from the meaning of lower-order representations that are their object. The elements of the meta-cognitive structures they represent cannot be related directly to the objects of the mental models they reflect. Consequently, the relations between them which constitute a higher-order mental model cannot be empirically falsified so that they appear, in spite of being constructed on the basis of empirical experiences, as knowledge a priori.

A particular consequence of this independence is that the symbolic transformations of a higher-order external representation need no longer to have any meaning at the level of the first-order representation of the real objects and actions from which the construction process started. This consideration provides an explanation for one of the most peculiar phenomena to be found in the individual as well as in the historical development of mathematical thought: in the course of development, formal conclusions contradict more and more interpretations of mathematical concepts at a lower level of abstraction, until finally, at a sufficiently high level of abstraction, those contradictions become irrelevant. In light of the theory of the function of external representations presented here, this phenomenon appears as an inevitable consequence of reflective abstractions.

A well known example is provided by the continuous expansion of the concept of apparently "natural" number by the development of seemingly unreal constructs such as "negative", "irrational", "transcendental" and "imaginary" numbers. Those numbers can no longer be interpreted as numbers in the sense of the original representation of sets of objects by counters. As long as numbers are applied exclusively to the context of their origin in transformations performed with concrete sets of discrete objects, such numbers therefore appear to be artificial constructions without any real correlate. Once numbers are understood formally, that is, as implicitly defined entities, however, the unreal character of these objects disappears.

A similar example is provided by non-Euclidean geometries. Those geometries appear absurd and unthinkable as long as geometrical concepts are, as Euclid's problems indicate, applied to finite figures constructed with compass and ruler. This application of the concepts makes the construction of Euclidean models of non-Euclidean geometries,[20] completely familiar to mathematicians and theoretical physicists and astronomers today, appear as unmotivated reinterpretations of the basic concepts of Euclidean geometry. What is used as a convincing technique of proof at a meta-cognitive level, appears to common sense as a violation of the seemingly "natural" meaning of geometrical concepts.

HISTORICAL STAGES OF THE DEVELOPMENT OF THE NUMBER CONCEPT

Based on the principles of a historical epistemology of logico-mathematical thought developed in the first part of this paper, we will draw in the second part consequences concerning the development of the number concept. After the description of basic arithmetical activities, stages of development will be defined that result from these activities through reflective abstraction.

Arithmetical activities as a basis of the concept of number

The arithmetical activities that have to be considered as a possible basis for the historical origin and development of numbers are closely related to fundamental structures of the number concept, that is, to the structures of the so-called "natural" numbers, the positive integers. First, these numbers are ordered in a linear sequence. Their use as ordinal numbers, that is, as a tool to identify orders, is based on this order. Second, these numbers can be interpreted as equivalence classes of finite quantities of discrete objects. They can therefore be used as cardinal numbers for the identification of quantities. Third, they exhibit two fundamental arithmetical operations, addition and multiplication. It is therefore possible to perform calculations with these numbers.[21]

In order to be developed, these fundamental structures require certain systems of actions. To determine relations of order that are to be represented by numbers, comparisons have to be made. To determine quantities, one-to-one correspondences have to be constructed. The additive relationship between three numbers results from to the combination of two sets of objects to form a third; the multiplicative relationship results from a repetition of the same action or from a reproduction of the same configuration of objects.

These actions of comparison, correspondence, combination and repetition, integrated into systems, will here be designated as *arithmetical activities*. They link the mental models of arithmetical thought to real objects and situations, thus constituting the basis for attributing quantitative values and relations to objects of empirical experiences. Their importance for the emergence of fundamental structures of numbers lies in the fact that they can be reasonably used independent of each other and, in particular, independent of any integration by a number concept. On the other hand, the symbolic representation of the results of performing comparisons between real objects, the construction of correspondences between them, and the realization of combinations and repetitions of them immediately leads to first-order representations which may guide their meaningful integration by reflective abstractions and the second-order representation of them in symbolic numerical systems.

Developmental stages of the number concept as meta-cognitive levels of reflection

The question of when, where and how such abstraction processes were realized in human history and what the actual outcome was is a question of historical research, but the general patterns such historical developments follow are determined by the nature of the underlying processes of reflection. At the outset any development of arithmetical activities is to a considerable degree based on cognitive universals independent of culture (Langer, 1980, 1986). Given the great cultural differences in the actual formation of even such fundamental arithmetical activities it is, however, obvious that even such fundamental arithmetical activities are already of a historical nature. They are not based on cognitive universals alone, but are also constituted by certain culture-historical processes of transmission. If developmental stages of arithmetical thought are therefore defined as reflective abstractions of historically developing arithmetical activities, that is, as meta-cognitive levels reflecting actions of comparison, correspondence, combination and repetition, this definition does not anticipate the answer to the question concerning the degree to which the concept of number is determined by cognitive universals. Rather the definition provides an analytical tool to study the question in historical sources.

If the historical development is interpreted as a sequence of gradually attained levels of abstraction reflecting arithmetical activities, the construction process must proceed in the following way (Damerow, 1996). At first, basic cognitive constructs of arithmetical thought are abstracted from the actions of arithmetical activities themselves. At the next level, they are abstracted from symbolic actions with symbol systems such as tally systems, body-counting, calculi, etc., which replace the original objects of the arithmetical activities. At higher levels, they are finally abstracted from the formal use of the written representation of logical conclusions concerning properties of numbers and numerical relations. As a result of this consideration, four main phases in the historical development of the number concept can be distinguished.

Before the development of the concept of number, there must have been a period characterized by the complete lack of arithmetical activities in the above defined sense. The emergence of the the number concept starts with the development of arithmetical activities as part of the culturally transmitted techniques. Their symbolic representation resulted in first-order representations of these arithmetical activities in the form of concrete tools for the control of quantities. A third phase is reached when cognitive constructs that originated from the reflection of real and symbolic actions of the first stage were represented by symbols, and culturally transmitted by means of those representations. Finally, in a fourth phase, the mental models which emerge from performing transformations with such symbols and the mental operations which constitute these models are made explicit and written down. In this way, they become formal rules for logical transformations coded in written language.

These phases can be interpreted as levels of reflective abstraction. Since the degree of abstraction is not determined by the process of reflection, the last two levels can be further subdivided according to this degree. The six resulting stages of reflective abstraction of the concept of number from arithmetical activities will be designated as:

stage 0:   pre-arithmetical quantification
stage 1:   proto-arithmetic
stage 2a: symbol-based arithmetic with context-dependent symbol systems
stage 2b: symbol-based arithmetic with abstract symbol systems
stage 3a: theoretical arithmetic with deductions in natural language
stage 3b: theoretical arithmetic with formal deductions

These theoretically postulated stages have to be validated by specifying their definitions and by identifying them historically, relating them to results of historical research in order to test the assumptions they are based on. For this purpose each stage will *first* be defined theoretically. *Second*, semiotic characteristics will be described that may serve as criteria for assigning arithmetical techniques of a cultural context to this particular stage of development. *Third*, some concrete historical examples will in each case illustrate the respective goals of research that result from the proposed theoretical model for a historical epistemology of the development of the concept of number.

Pre-arithmetical quantification (stage 0)

Definition: The level of pre-arithmetical quantification is a stage of development in which comparisons are the only arithmetical activities. Pre-arithmetical quantifications are based only on comparisons. They include neither the construction of correspondences as, for instance, by counting, nor the composition of quantities by arithmetical operations, for instance the construction of numbers through the repetition of units.

Semiotic characteristics of pre-arithmetical quantification. The most noticeable difference of the level of pre-arithmetical quantification from all levels of genuine arithmetical activities is the absence of socially transmitted standards that might serve as tools for the construction of one-to-one correspondences. On the pre-arithmetical level there are no structured counting sequences and no tally systems such as finger counting, counting notches, counting knots or calculi. No words, signs or other symbols are used that possess any arithmetical meaning. The language at this stage possesses terms for quantities, yet these are of exclusively qualitative nature. Insofar as there are any rudimentary words for numbers at all, these are not used for counting; rather, they are designation for special qualities, that is, designations for intuitively and holistically understood small quantities. The quantitative aspects of an object of cognition are not yet distinguished from its specific physical appearance and from implications of its quantitative aspects (Ferreira, 1997).

Historical identification of pre-arithmetical quantification. A historiogenetic theory of arithmetical thought has first to address the question as to which original conditions for the historical development of this thought pattern are already preconditioned by cognitive universals founded in human nature and therefore not subject to historical change. In particular, the question arises as to whether cultures ever existed that correspond to the definition of a pre-arithmetical level given here. The answer to this question ensues from the fact that non-literate cultures existed until recently, which used no counting techniques before their contact with Western cultural traditions. The definition of a pre-arithmetical level applies to such cultures. The best known examples are the Australian aborigines (Dixon, 1980: 107f; Blake, 1981: 33; numerous examples in Dixon & Blake: 1979ff) and the South American natives (Lévy-Bruhl, 1926: 181-184; Gnerre, 1986: 74). It is, however, difficult to identify such cultures with certainty today since even the most remote primitive peoples have been in extended contact with modern civilization. Through trade, which often provided the first systematic contacts with Western culture, they quickly assimilated arithmetical activities and concepts and also

33

changed quantifications in their own language, for example, by inventing new, indigenous words for counting and new arithmetical techniques (Dixon, 1986: 108; further, in particular, Saxe, 1982). Proof that many of these cultures did not possess developed counting procedures before their contact with Western culture is today only feasible through often speculative linguistic inferences. Moreover, the linguistic material can often only be collected by interviewing a few elderly informants,[22] whose children and grandchildren attend public schools in order to learn reading, writing and arithmetic.[23] The influence of such contacts in many cases invalidates the information that can be gained from the informants, in particular because it consists primarily in the adoption of arithmetical activities which simply change the semantics of existing terms.

Identifying with certainty the pre-arithmetic level in the development of arithmetical thought *historically* constitutes an even larger problem, since written sources usually do not reach far enough into the past even to gain linguistic material for an identification according to semiotic criteria. An important clue is provided by the fact that from periods before the Late Neolithic no objects or signs, for example counting notches or calculi, have been identified that might have served as tally systems, or might have had another kind of arithmetical function. It is true that Palaeolithic, Mesolithic and in particular Neolithic finds, especially of bones, often exhibit repeated signs such as regular patterns of notches, and that these have occasionally been interpreted to be representations of numbers,[24] but such an interpretation can hardly be justified, since these sign repetitions lack the characteristic subdivision by counting levels that would be expected in signs with numerical meaning, and which is indeed present in all known real counting systems.

Proto-arithmetic (stage 1)

Definition: The level of proto-arithmetic is a stage of development in which first-order representations of quantities are constructed by means of one-to-one correspondences to standard sets of concrete objects or other symbols.

Semiotic characteristics of proto-arithmetic. The earliest genuine arithmetical activities historically attested are without exception based on objects themselves being represented by symbols, their quantity however by the repetition of these symbols. Symbols are the most simple tools for the construction of one-to-one correspondences that can be transmitted from generation to generation. Structured and standardized systems of symbols, from which standard amounts can be formed that are assigned to the quantities to be identified, are the oldest tools for the identification and control of quantities.

Since symbols, in order to represent quantities, may be repeated either temporally or spatially, in principle two different kinds of such simple standards for the representation of quantities can be distinguished. The former will here be called counting sequences, the latter tallies.

A *counting sequence* in this sense is a standardized sequence of words or symbolic actions which is assigned to the elements of a given set in a fixed sequence, realized in time, a process that is generally called counting. Through the process of counting, a linear (temporal) first-order representation of the ordinal structure of such a given quantity is realized.

Tallies, on the other hand, are concrete objects such as signs, knots, notches or calculi, that can be arranged and combined in a simple way for the purpose of constructing correspondences. By assigning such objects to the elements of a given set, a spatial first-order representation of the cardinal structure of a given quantity is realized.

Thus, counting sequences and tallying systems are two different forms of first-order representations of finite sets of objects, representing different aspects of their quantity. Common to both forms are certain characteristic structures originating from their function. The deliberate construction of correspondences initiates, in the course of historical development, processes acting to extend their range of application. Genuine counting sequences and tallying systems therefore exhibit two typical structural patterns that are a consequence of their progressive improvement in the direction of an infinite counting sequence. First, older counting limits are preserved as inherent steps of the counting procedure. Second, counting limits are systematically overcome by certain techniques.

The predominant procedure for extending the range of application of counting sequences and tallying systems is, when a counting limit is reached, to start over with counting in connection with a counting procedure of higher-order which determines how often the primary procedure had been used. The repeated application of such techniques of passing counting limits generates a hierarchically structured symbol system which can, with this technique, be extended to virtually any required order of magnitude.

The characteristic structures of genuine counting sequences and tallying systems resulting from this procedure are important indicators of arithmetical activities that have to be attributed to the proto-arithmetic level. On the basis of their hierarchically organized structures, tools for the construction of one-to-one correspondences that have served in the control of quantities can be identified even when there are no historical sources that give us definite information concerning their original purpose. Thus there is in principle no problem in distinguishing cultures that have reached the proto-arithmetic level from cultures at the pre-arithmetic level.

The proto-arithmetic level is distinguished from higher-order levels through the absence of arithmetical procedures, that is, of symbolic transformations, that correspond to arithmetical operations like addition and multiplication. Such arithmetical procedures require that the symbols which are transformed according to formal rules no longer represent the counted objects, but rather their quantity. According to the definition of the proto-arithmetic level this level is precisely characterized by the fact that symbolic transformations apply to representations of objects and not to the representation of representations characteristic of higher-order levels (Lévy-Bruhl, 1926: 181-223; Gay & Cole, 1967; Hallpike, 1979: in particular 236-279).

Historical identification of proto-arithmetic. Proto-arithmetical tools and techniques are known to us mainly from surviving non-literate cultures. Such cultures are to be assigned to the proto-arithmetic level if they actually developed arithmetical activities. They display a wide variety of counting sequences and tallying systems for the construction of correspondences, realized in all kinds of forms, whereas arithmetical techniques that are based on the symbolic representation of quantities and the numerical relations between them are encountered relatively seldom. Their proto-arithmetical tools are used almost exclusively for the identification of quantities, not, however, for symbolic transformations with the purpose of quantitative prediction of results from real interaction with sets of objects.

The study of the proto-arithmetic level in surviving non-literate cultures is of great significance for the reconstruction of the development of the concept of number insofar as there is almost no opportunity to study this stage of development in historical sources. The study of such surviving cultures provides hints for the interpretation of pre-literate period archaeological finds with possible arithmetical functions. Comparisons of various surviving non-literate cultures with regard to their stage of development in arithmetical thought suggest, in particular, that agricultural cultivation, animal husbandry and housekeeping connected with sedentariness provide social conditions that make proto-arithmetical techniques useful and their systematic transmission and development possible. It thus turns out to be likely that the historical process of cultural development arrived at the level of proto-arithmetic in the Late Neolithic and the Early Bronze Age.

A system of clay tokens apparently possessing proto-arithmetical functions has indeed been identified as one widely used in the Near East during that period,[25] that is the time span from the beginning of sedentariness in the areas surrounding the Mesopotamian lowland plain and in the Nile valley around 8000 B.C. until the emergence of cities around 4000 B.C. Thousands of these tokens were found in excavations, especially in those of the Mesopotamian alluvial plain and the Persian highland. The oldest clay objects ascribed to these symbols are dated to the beginning of the 8th millennium (Schmandt-Besserat, 1992: 36f). Their identification as belonging to a tallying system is based on the fact that in the transitional phase to symbol-based arithmetic they were converted into in the numerical signs of the proto-cuneiform writing system developed around 3200 B.C.

The transition from proto-arithmetic to symbol-based arithmetic

Although surviving non-literate cultures provide detailed knowledge about proto-arithmetic techniques, the conclusions that can be drawn concerning archeological finds from historical time periods of proto-arithmetic are very limited. The first phase in the development of arithmetical thought about which we have detailed information provided by historical sources from different geographical areas is the subsequent phase of symbol-based arithmetic. One of these areas, however, offers in addition important information for the understanding of the origin of arithmetic.

Recent studies of archaeological sources from the dawn of literacy in the Near East have lead to the identification of peculiar forms of arithmetical activities which are, in all likelihood, phenomena of the transition from proto-arithmetic to symbol-based arithmetic which may be defined as an intermediate phase.

Definition: The transitional phase between proto-arithmetic and symbol-based arithmetic is a historical phase during which, just as at the proto-arithmetic level, exclusively first-order representations of quantities are constructed. These are, however, in contrast to the proto-arithmetic level, represented in a developed symbol system and not only by arithmetical activities such as counting and tallying.

Semiotic characteristics of the transitional phase. The characteristic semiotic feature of this transitional phase are complex symbol systems used as counting units, whose numerical values, however, are not constant, but vary with the context of their application. They do not represent context-independent fixed numerical meanings, but rather units of counting and measurement of products whose numerical relations are determined by the social context in which they are standardized by conventions. They differ in particular from the counting and tallying systems of the proto-arithmetic level in that they are already the subject of genuinely symbolic transformations. The basis is formed by transformations that still are first-order representations of real actions; that means they have to be interpreted as representations of economic transactions or real administrative activities. In addition, transformations occur that are not only symbolic representations of real actions, but use the potential of symbol systems for performing transformations with symbols the corresponding real actions of which are useless or even impossible. The purpose of such transformations is merely to get knowledge about their outcome.

Historical identification of the transitional phase. The transition from proto-arithmetic to symbol-based arithmetic is obviously related to the invention of writing. However, cuneiform writing is presently the only writing system with rich archaeological findings reaching back to the period of the transition from proto-arithmetic to symbol-based arithmetic. These findings include thousands of clay tokens, numerous numerical tablets, and about 7000 proto-cuneiform and proto-Elamite texts and text fragments, almost exclusively economical texts with records of quantities exhibiting the semiotic characteristics described above.[26]

The transition from proto-arithmetic to symbol-based arithmetic occurred during a relatively short period of time. The starting-point consisted of a system of clay tokens which was in use as a proto-arithmetical tool in the Neolithic period of the entire Near East. In all likelihood, these tokens served as first-order representations of counting and measuring units, probably constituting an administrative control system for goods and products. Together with the emergence of Mesopotamian cities and the beginnings of a form of state organization and the centralization in the administration of estates during the second half of the 4th millennium, this administrative control system was fundamentally revised. The tokens were first complemented and later completely replaced by markings that were impressed onto the surface of sealed clay tablets, the so-called numerical tablets. Around 3200

B.C., the numerical notations on these tablets were supplemented with proto-cuneiform pictograms, thus constituting the so-called proto-cuneiform writing system of Mesopotamia. The tablets also achieved a more complex structure by combining several quantitative notations on one tablet, arranged according to their function. Moreover, such numerical entries were often totalled. The number of numerical signs increased to some 60 signs, each representing units of measurement and counting.

These numerical signs exhibit the peculiar semiotic property which suggest the definition of the transitional phase given here. Unlike all other developed systems of numerical notations known so far, most of the proto-cuneiform numerical signs were used to represent more than just one unit of counting or measurement, regardless of the quantities represented by these units, i.e., the signs had no fixed but context-dependent numerical values. For the whole period of some 200 years during which the proto-cuneiform writing system was used the sources show no attempt to attain unambiguous numerical values for these signs.[27] Already the texts of the Early Dynastic period, the first period following the time of proto-cuneiform writing, exhibit, however, the typical semantic structures of the numerical signs of symbol-based arithmetic and show scarcely any traces of the peculiar semantics of numerical signs from the transitional period.

Symbol-based arithmetic (stage 2)

Definition: The level of symbol-based arithmetic is a stage of development in which second-order representations of quantities and arithmetical activities are constructed by the reflection of proto-arithmetical mental models and by their representation in a symbol system. This development produces systems of numerical signs, which are semiotically structured according to the mental models they represent. The level of symbol-based arithmetic can be further subdivided with regard to the degree of abstraction from the specific context in which the meaning of the symbols is generated by reflection, resulting in a sublevel of context-dependent symbol systems and a sublevel of abstract symbol systems.

Semiotic characteristics of symbol-based arithmetic. The striking characteristic of the level of symbol-based arithmetic is the emergence of complex systems of numerical symbols and of formal rules for their application. Two different forms of such systems fulfilling different purposes can be distinguished, namely numerical sign systems and calculation aids. The former are used predominantly for the recording of quantitative information, the latter for its processing.

The formal rules for dealing with numerical sign systems and calculation aids may relate directly to the symbols, and may thus be explicitly stated. They may, on the other hand, only emerge from the strict application of the symbols according to their formal meaning. Such implicit rules can be identified by the consequences of their application, because their application is strict in the sense that their applicability is no longer dependent on contingent conditions of the actual application context in which they are transmitted.

At the level of symbol-based arithmetic, the formal rules of symbol transformations usually remain implicit. Although at the level of symbol-based arithmetic usually the rules for symbolic transformations are still only implicit rules which can, moreover, be of a specific nature for their respective different contexts of application, they represent the first genuine *arithmetical techniques*. They constitute a new level of arithmetical activities applied to objects which are no longer the original real objects of the area of application, but the symbols of numerical notation systems and calculation aids and their abstract meanings. Therefore, at this level the direct relationship of counting sequences and tallying systems to the real objects they represent becomes more and more obsolete and the repetition of symbols in first-order representations gradually loses its function of supporting the arithmetical activities. Numerical notations as second-order representations are reduced to standardized signs used as names for entities in a mental model.

Context-dependent and abstract symbol systems. Since numerical symbol systems characteristic of the level of symbol-based arithmetic are, due to the implicitness of their semiotic rules, at first still closely related to the meanings of the symbols in specific contexts of application, they are not necessarily from the outset applicable to arbitrary contents.

At the *sublevel of context-dependent symbol systems* quantities are already symbolically represented by second-order representations, but they still possess specific areas of application. A generalized system of notations into which all special forms of notations can be transformed and thus become standardized is still lacking. The rules of sign transformation remain intermingled with the specific meanings of the signs in their particular context of application and thus are rarely formal rules, that is, rules depending only on the form of the signs and sign combinations.

At the *sublevel of abstract symbol systems* there exists an arithmetical symbol system that is generally applicable, independent of the context of application. It allows the standardized representation of all quantities, into which all particular application-specific representations can be transformed. There are no external limits to its application. Consequently, with such symbol systems formal operations can be performed, that is, without reference to any specific interpretation in their actual context of application.

Such abstract semiotic constructs without canonical reference to specific contexts of real objects and actions make possible for the first time forms of cognitive processing of abstract ideas that may be interpreted as early forms of genuine mathematical thought. The constitution of meanings reflecting formal symbolic transformations leads to knowledge of entirely implicitly defined artificial objects, with at best metaphorical reference to real contexts of application. This knowledge can be acquired, represented and historically transmitted as a first body of mathematical knowledge. However, the knowledge of abstract objects that are only mentally constructible is at this level not yet integrated into deductive systems with formal rules of inference. Thus the qualities of these abstract objects cannot yet be derived or substantiated with logical derivations. But mental operations with such objects created by reflection are for good reason usually considered an early form of mathematical thought.[28]

Historical identification of symbol-based arithmetic. Most, if not all, advanced civilizations, in particular the Egyptian empire, the Mesopotamian city states, the Mediterranean cultures, the Chinese empire, the Central American cultures and the Inca culture, have, independent from each other, developed or adapted from other cultures systems of numerical symbols that exhibit the semiotic characteristics of the level of symbol-based arithmetic. The basic symbols of these systems were, the same as at the proto-arithmetical level, signs for units and not for numbers, but part of the systems were now complex symbol transformations, that is, arithmetical techniques such as the Egyptian calculation using unit fractions (Neugebauer, 1926: 137ff; Chace, 1927; Vogel, 1958-1959, vol. 1: 31-44; Gillings, 1972: 20ff; Damerow, 1995: 176-199), the sexagesimal arithmetical technique of the Babylonians (Neugebauer, 1934: 4ff; Vogel, 1958-1959, vol. 2: 15-35; Damerow, 1995: 199-261), the transformations of rod numerals on the Chinese counting board (Needham, 1959; Juschkewitsch, 1966: 12ff; Li & Dù, 1987: 3-19), the calendar calculations in the pre-Columbian culture of the Maya (Thompson, 1941, 1960: 51ff; Gaida & Tear, 1984; Closs, 1986; Damerow & Schmidt, 2004: 143-145, 163-167), or the technique of the use of knotted cords (quipú) as administrative tools by the Inca (Locke, 1923; Ascher & Ascher, 1971-1972; Scharlau & Münzel, 1986: 80-93).

In all of these cultures symbol-based arithmetic emerged in a context-dependent form. Administrative bureaucracies have always been the institutions that created those complex techniques for the transformation of numerical symbols characteristic of the level of symbol-based arithmetic, and the specific purposes for which these techniques were developed determined the structure of the symbolic transformations. Consequently, the early systems of symbol-based arithmetic exhibit a great variety of formal structures. With the exception of addition, which, as a trivial consequence of the representation of quantities by repetition of symbols, developed in all cultures in almost the same form, the arithmetical techniques of the early civilizations reflect for the most part the culture-specific differences of the areas of application for which the respective systems had been developed.

Historical identification of the transition from context-dependent to abstract symbol systems. The historical identification of the transition from the exclusive use of context-dependent symbol systems to the construction of a unified abstract symbol system raises considerable difficulties. Only the actual use of a seemingly abstract symbol system shows whether it serves as a tool for rendering different forms of context-dependent numerical notations in a unified form of representation. But the sources preserved from early civilizations are often too meagre to allow of a sufficiently precise assessment of the area of application of a symbol system that would make it possible to distinguish between a context-dependent and an abstract use of such a system. Moreover, numerical symbol systems exhibit such great differences with regard to the scale of their area of application that some, when compared to the others, appear to be much more abstract systems. These are mostly the counting sequences, for counting procedures in contrast to measuring procedures are as a rule so object-neutral that fortuitous peculiarities of the counted objects can scarcely influence the counting procedure. They can therefore easily be misunderstood as indications of an abstract arithmetic which, in fact, may not yet have existed.

The distinction between a sublevel of context-specific numerical symbol systems and a sublevel of abstract systems has been proposed here in particular in view of the historical development of arithmetic in Babylonia. Since clay used in Mesopotamia as writing material is extremely durable an abundance of administrative and mathematical cuneiform texts survived which document the development of arithmetical techniques from the earliest beginnings of proto-arithmetic to the creation of an abstract symbol system of symbol-based arithmetic.

These sources demonstrate that from the moment of the emergence of writing around 3200 B.C. until the invention of the sexagesimal place value system about 2000 B.C., exclusively context-specific symbol systems were used (Damerow, 1995: 219-237). One of these systems was a sexagesimal system.[29] It was already strictly structured at the moment of the emergence of writing, probably corresponding to a similar sequence of number words.[30] This sexagesimal system had already a wide range of application. It was, however, not used for all objects. It range of application was essentially restricted to counting discrete objects.

Only with the invention of the sexagesimal place value system around 2000 B.C. was an abstract system of numerical notations introduced that unified all forms of notations.[31] Some 100,000 administrative documents survived from the last 100 years of the third millennium B.C., the period of the third dynasty of Ur, which provide detailed information about the accidental circumstances which triggered this invention and about the systematic tendency towards a unification of numerical notations as well. This tendency in combination with specific circumstances was the ultimate cause of the transition from context-dependent to abstract symbol-based arithmetic in Mesopotamia.

The historical context which provided the conditions for the invention of the sexagesimal place value system was the unification of the city states by the empire of the 3rd Dynasty of Ur and the building up of a centralized administration of all resources, products and services. Within this context highly specialized book-keeping procedures with specialized metrologies and corresponding numerical notation systems had to be unified in order to get standardized judgements about values as a precondition of the universal exchange of such resources. At first, different administrative units, depending on the focus of their tasks, used specific value standards into which all resources were converted, standards such as quantities of silver, fish, barley or human labour (Englund, 1990: in particular 18ff, 96ff, and 181ff). Finally, however, the abstract sexagesimal place value system was developed which had no longer any association with specific metrologies. Metrological tables were used to convert the traditional notations into notations of this new system.

The invention of the sexagesimal place value system in Babylonia had two far-reaching consequences for the development of arithmetic. First, it revolutionized the Babylonian arithmetical technique. The multitude of arithmetical procedures developed for the solution of specific problems was replace by one set of algorithms, including in particular a uniform multiplication procedure (Damerow, 1995: in particular 246-258). Second, the invention of the sexagesimal place value system had the development of the so-called Babylonian mathematics as a consequence. Reflection on the arithmetical operations of the sexagesimal place value system formed the basis for the construction of a system of technical terms, canonical types of problems and abstract mental operations which was only metaphorically related to its areas of application. Operations with the sexagesimal place value system made it possible to solve such complex, but at the same time practically irrelevant problems as the computation of the sides of a field from its perimeter and its area, that is of the second-degree equation of identifying two unknown quantities from their sum and their product.[32]

This description of the process and the consequences of the transition from context-dependent to abstract symbol-based arithmetic is based on the exceptional case of Babylonia where an abundance of sources survived documenting this transition. It is, however, likely that the symbol-based arithmetic developed independently also in other cultures such that the ancient Egyptian and the ancient Chinese arithmetic developed in a similar way from context-dependent systems.

Concept-based arithmetic (stage 3)

Definition: The level of concept-based arithmetic is a stage of development in which second and higher-order representations of symbolic actions with representations of symbol-based arithmetic are constructed by the reflection of mental models depending on symbol-based arithmetic and by representing these meta-cognitive models in the medium of written language. This process of reflection results in logically structured systems of arithmetical propositions that make their deductive derivation and proof possible. The cognitive constructs that are thus abstracted from sign systems of symbol-based arithmetic become progressively independent of their specific properties. The level of concept-based arithmetic can be further subdivided with regard to the degree of abstraction from the specific context in which the meaning of the sign systems is generated by reflection, resulting in a sublevel of deduction in natural language and a sublevel of formal deduction.

Semiotic characteristics of concept-based arithmetic. The most important semiotic characteristic of the level of concept-based arithmetic is the explication of general propositions concerning the properties of abstract numbers. Since such propositions can only be deduced, they are naturally embedded in a system of deductive relations which connect them with each other. Under certain conditions such deductive webs of relations can be linearised, that is, the propositions can be globally structured in a way that all circular arguments are removed and all propositions appear to be systematically deduced from but a few basic propositions. Accordingly, it is customary following the Euclidean tradition to write such propositions down in a deductive order as a *theory*.

Theoretical concepts do not owe their structure directly to the arithmetical techniques that constitute them, but to the knowledge that can be gained by the application of those techniques. These concepts are consequently embedded in structures of argumentation, that is, in structures at a meta-level of reflection. At this level of reflection, the meanings of arithmetical concepts that were seemingly determined by the rules of symbol-based arithmetic, may again be subject to development and can, if necessary, be modified by considerations of suitability of a higher kind. The concept of *prime number* as it was handed down to us by book VII of Euclid's *Elements* (Heath, 1956, vol. 2: 296-344), for example, does not have the same kind of immediate technical meaning as do technical terms of symbol-based arithmetic such as sum, factor or divisor. Only the reflection of all possible summations, multiplications, divisions, and their results that can be inferred in a deductive system, leads to the formation of such a concept as a prime number being a number that cannot be further decomposed into factors other than one and the number itself and to the derivation of propositions such as the proposition that every natural number can be unambiguously factored into a product of prime numbers.

The meanings of such concepts resulting from the reflection on mental models depending on symbol-based arithmetic differ considerably from the meaning of concepts based on the symbol-based arithmetic itself. While the latter get a justification from their practical applicability, the former are apparently independent of any applicability, seemingly being determined only by internal consistency of deductive relations. Abstract numbers appear to exist *a priori*, because they have their origin in reflection. Numerical notations that are used to handle them for practical purposes, for example, appear to have no influence on the truth of statements about their properties. On the other hand, new constructions become possible on a meta-level of reflection. Questions such as the question of whether numbers might be conceivable with properties differing from what can be derived under the given conditions turn out to be justified even if such imagined numbers have no meaningful interpretation any longer in the context of arithmetical activities from which the process of mental construction started.

Deduction in natural language and formal deduction. Similar to the level of symbol-based arithmetic, the level of concept-based arithmetic can, according to the degree of abstraction, be divided into two sublevels.

The *sublevel of deduction in natural language* is characterized by the fact that the deductive systems consist of statements and proofs that are formulated in natural language. Mathematical terms, for instance the concept of number, are explicitly defined and are abstract insofar as they are determined within the logical structure of a deductive system. They still refer, however, to concrete objects and actions, since a representation in natural language entails connotations of the concepts that are determined by their origin in real actions. Numbers, for example, have an abstract structure on the one hand, on the other a connection to the arithmetical activities that the numerical notations are based on; they not only have provable also apply to quantities of real objects.

Arithmetical concepts at this sublevel of development thus have at the same time *extrinsic meanings* that result from their origin in arithmetical activities and *intrinsic meanings* that are deduced from seemingly self-evident axioms. This results in particular in a canonical meaning of the concept of number at this level of development, usually expressed by the term *natural number*. Natural numbers are extrinsically determined as reflectively constructed structures of actions of correspondence and comparison. As cardinal numbers of quantities, they have thus a canonical object that determines their quasi natural properties. Intrinsically, their properties are determined by universal laws such as the distributive law or the commutative law of addition and multiplication. Those laws can be arranged deductively in a way that makes them appear to be logical conclusions of a few, seemingly self-evident axioms, for example the Peano axioms.

At the *sublevel of formal deduction*, the concepts formulated in natural language are replaced by terms of formal languages, so that connotations with their original meanings are systematically avoided. The concepts can be subsumed under generalized, unifying concepts that may be constructed artificially, and be entirely determined by mathematical structures. Their meanings no longer appear to be determined naturally, but through axioms that seem to be presupposed arbitrarily.

Numbers at this level appear as superposition of algebraic and topological structures precisely defined by axioms. Such axioms can be modified and in various ways combined with each other into suitable new structures designated artificially as semigroups, groups, topological groups, rings, fields, etc. Numbers have no longer an exceptional status among these mathematical objects; their apparent "naturalness", characteristic for the sublevel of deduction in natural language, appears to be an fortuitous historical relict.

Historical identification of concept-based arithmetic. According to the definition of the level of concept-based arithmetic, the transition to this level is not a process specific to arithmetic. It rather applies to the reflection of mental models representing symbolic actions in general, provided that the results are encoded in the medium of written language. The encoding is essential for the transition to a concept-based cognitive construct. The reflection of mental models representing symbolic actions is in itself an inherent outcome of any symbol-based activity. In the case of symbol-based arithmetic such processes of reflection led to the emergence of various forms of so-called pre-Greek mathematics in early civilizations such as ancient Egypt (Neugebauer, 1934; Gillings, 1972), Babylonia (Neugebauer, 1934; Friberg, 1990; Høyrup, 2002), India (Joseph, 1992), and China (Li & Dù, 1987). Such systems were transmitted within a culture from one generation to the next primarily by fostering the building up of the culture-specific mental models through exercising the symbol-based activities which they represent, partly also by teaching explicit rules of how to solve problems by means of symbol-based arithmetic. However, the first examples of an explicit representation of chains of inferences, which are based on abstract mathematical objects and result in universal propositions about such mental entities, are known from ancient Greece.[33]

The starting point of the development of Greek mathematics was the explicit representation of properties of numbers implicitly embodied in abstract numerical symbol systems in the medium of natural language (Szabó, 1960, 1978, 1994). This medium at the same time preserves the relation of numerical concepts to their origins in arithmetical activities. Thus in the Greek tradition numbers appear to be logically determined and at the same time empirically valid with regard to the objects of these activities; this is made explicit in the Platonic concept of pre-existing ideas.

The oldest known example of a deductively ordered system of propositions concerning properties of abstract numbers is the so-called doctrine of even and odd numbers, a theory that has not come down to us in its original form through preserved sources, but which can be reconstructed from definitions and theorems of Euclid's *Elements*.[34] The theory goes back to the Pythagoreans of the 5th century B.C., who tried to associate all objects with numbers (Becker, 1936; van der Waerden, 1947/1949, 1954: 96-97, 108-127; Heath, 1921, vol. 1: 67). Its propositions are mainly concerned with the dependence of the property of a calculated number to be either even or odd on the respective properties of the original numbers entered into the calculation. The 14 relevant theorems that have come down to us reflect experiences with geometrical configurations of Greek counters, the so-called figured numbers (van der Waerden, 1954: 98-100). In the modified form in which the theory has been transmitted by Euclid, the theorems appear to be arranged in a deductive schema (Lefèvre, 1981; Damerow & Schmidt, 2004: 167-173). This development of the theory from the symbolic representation of numbers by arrangements of counters to a representation in written language and deduction of their properties from basic assumptions marks precisely the transition from symbol-based to concept-based arithmetic in the sense of the definition proposed here.

Euclid (Heath, 1956, vol. 2: 277, 280) transmits in his *Elements* also a more comprehensive theory of numbers, based on the Platonic[35] definition "number is a quantity composed of units." But even this theory could cover only a small part of the range of existing symbolic actions of symbol-based arithmetic provided by pre-Greek mathematics. The concepts of Greek theoretical arithmetic compiled in Euclid's *Elements* were incompatible with arithmetical activities which did not belong to its foundations. In particular, fractions could not be related to the theory of number based on counting sequences and tallying systems represented by the definition "number is a quantity composed of units," from which the properties of numbers were deduced. Platonism dogmatically excluded from theoretical arithmetic all numerical structures incompatible with this definition.[36] For a long time to come, concept-based arithmetic coexisted with the much richer tradition of symbol-based arithmetic.

Historical identification of the transition to formal deduction. The further development of symbolic means to represent propositions derived from symbolic actions of symbol-based arithmetic produced formal structures of higher-order levels of reflection, structures which were increasingly distant from the basic arithmetical activities that underlay the Greek theory of numbers. Historically, this process presents itself as a continuous process of the construction of numbers by reflective abstractions which did not fit the ancient definition of number. As long as the ancient concept of number determined thinking, they were perceived as "absurd numbers"[37], "irrational numbers" and "imaginary numbers," since they were incompatible with the historical connotation of the term number to designate attributes of concrete sets of objects consisting of enumerable units.

The precondition for overcoming this understanding of number and thus for the transition to the level of formal deduction was the development of the analytic tradition which provided techniques of designating "unknown" mathematical objects such as numbers by using letters, in modern terminology "variables," instead of meaningful descriptions by words, and operating with them as if they were the mathematical objects themselves. Such operations could be performed strictly according to the symbolic actions on higher level representations of symbolic arithmetic without any connotations with the original meaning of the mathematical objects they represent as, for instance, the Platonic definition of natural numbers.

Descartes and Leibniz were early advocates of the analytical method. This method initiated in the course of the 18th century the creation of completely new mathematical disciplines which for a long time competed with the "synthetic" method of mathematical disciplines using the Euclidean style of deductive mathematical reasoning. It was, however, not before the end of the end of the 19th century that both traditions merged into the modern technique of formal proofs using implicit definitions by systems of axioms to determine the nature of mathematical objects independent of meanings derived from their historical origins. All types of numbers were relegated to examples of such implicitly defined mathematical objects.

Final remarks

Historical epistemology is supposed to answer the question of whether in the history of knowledge systems stages of development can be identified which at each time determine the range of possible knowledge achievements in individual cognition. Accordingly, a historical epistemology of the development of the concept of number aims at reconstructing the changing structure of arithmetical thought which is represented in the diversity of arithmetical techniques studied by historians of science.

The result of an epistemological investigation certainly cannot consist of replacing historical arguments with theoretical arguments. Indeed, the theoretical considerations in the first part of the work have by no means made historical research redundant. The answer to the question, to what degree the structures and processes of arithmetical thought represent culture-independent and historically unalterable universals of the nature of *homo sapiens*, and what part of arithmetical thought on the other hand derives ultimately from cultural achievements and what structures of this cognition have developed in which historical periods, is by no means theoretically pre-empted by the considerations presented here. On the contrary, they provide a precondition to study and answer such questions by historical and cross-cultural studies, for the theory of the historical transformation of arithmetical knowledge by reflective abstractions presented here opens up an opportunity of reconstructing the nature of arithmetical thought under given historical conditions through an analysis of its external representations.

The brief outline of the historical development of arithmetic presented in the second part of the paper is thus not derived from the theory of the historical transformation of arithmetical knowledge presented in the first part, but rather the result of relating developmental stages of reflective abstraction to obvious differences between representations of arithmetical knowledge in different historical periods. This result can be briefly summarized in the following schema.

Stage 0   *Pre-arithmetical quantification*: approximately until the end of the Mesolithic period (in the Near East until ca. 10,000 B.C.).
No arithmetical activities. All judgments about quantities are based on direct comparisons of amounts and sizes. Communication and transmission only by transmittable techniques of comparison and by comparative expressions of language.

Stage 1   *Proto-arithmetic*: Neolithic period and Early Bronze Age (in the Near East until ca. 3000 B.C.).
Quantities are precisely identified by one-to-one correspondences. Communication and transmission with the aid of conventional counting sequences and tallying systems.

Stage 2a   *Symbol-based arithmetic with context-dependent symbol systems*: Period of the early city cultures (in the Near East until the invention of the sexagesimal place value system around 2000 B.C.). Quantities are structured by metrological systems. Communication and transmission of these systems and of the corresponding mental constructs through complex symbol systems and developed techniques for the transformation of symbol configurations.

Stage 2b   *Symbol-based arithmetic with context-independent symbol systems*: Period of developed city cultures (in the Near East until the beginning of Classical Antiquity around 500 B.C.).
Quantities are structured by abstract numerical systems with object-independent arithmetical operations. Communication and transmission of these systems by unified, context-independent, but culture-specific symbol systems for the representation of arbitrary quantities, including abstract "rules of calculation". Emergence of first forms of "pre-Greek mathematics" that are abstract but dependent on culture-specific symbol systems.

Stage 3a   *Concept-based arithmetic with deductions in natural language*: Classical Antiquity, Late Antiquity, Middle Ages and Early

Modern Era (until the emergence of analytical mathematics in the 18th century A.D.).

Abstract number concept with "a priori" provable properties. Communication and transmission with the aid of a written representation of "propositions" about abstract numbers and their mathematical properties. Propositions are logically ordered and systematically arranged by deductive theories according to the model of Euclid's *Elements*.

Stage 3b   *Concept-based arithmetic with formal deductions*: The modern mathematical tradition until the present.

Formal understanding of arithmetical structures and expansion of the number concept by construction of new arithmetical structures. Communication and transmission with the aid of formal language systems.

This schema is preliminary. It can be only the first step of realizing the aim of a historical epistemology of arithmetic to explain how arithmetic evolved from its roots in incipient arithmetical activities which survived in certain non-literate cultures to the abstract and highly complex arithmetical thought of modern mathematics. It provides, however, a theoretical framework which makes it possible to pursue historical research that reconstructs the development of arithmetical thought as an outcome of the changing material culture of calculation.

NOTES

[3] See the diverse arguments for this "Platonic" view, for example, in transcendental idealism, particularly Kant's arguments in his *Prolegomena* (Kant, 1985, §§6-13) and in his *Critique of pure reason* (Kant, 1998, A 154-158 = B 193-197: 281-283); in neo-Kantianism, particularly Cassirer's arguments in his *Philosophie der symbolischen Formen* (Cassirer, 1953-1957, vol. 1, ch. 3, sect. 3, in particular: 238f, and vol. 3, part 3, ch. 3: in particular 341-356); in logical positivism, for example Frege's arguments in *The foundations of arithmetic* (Frege, 1959), and Carnap's arguments in his *Foundations of logic and mathematics* (Carnap, 1955); further, in constructivism, Lorenzen's arguments in *Einführung in die operative Logik und Mathematik* (Lorenzen, 1955).

[4] See, for example, the historical accounts of Tropfke (1930-1940, 1980), Menninger (1970), Gericke (1970), and Ifrah (2000).

[5] Brainerd (1979: 3-22) at the beginning of his study on the origin of the concept of number discusses its historical development. His psychological remarks concerning the various historical periods, for example the assertion that an abstract concept of number is already apparent in Egyptian numerical notations from the time around 3500 B.C., remain, however, unsubstantiated and reflect no discernible connection with the remaining content of the study.

[6] In a number of studies such a parallelism of ontogenesis and historiogenesis is discussed from different viewpoints (Bachelard, 1994, 1996; Arcà, 1984; Strauss, 1988; Dux, 1992: in particular 23-35, 1994).

[7] A "mental model" (Gentner & Stevens, 1983) is conceived of here as special type of "frame" (Minsky, 1975, 1985; Davis, 1984) that represents shared, historically transmitted knowledge. On the nature of arithmetical skills see also Ashcraft (1992) and Campbell (1992).

[8] An intensive debate on culture-relativistic conceptions of cognition was provoked by Whorf (1956). On relativism in cognition see also Pinxten (1976) and Gumperz & Levinson (1996).

[9] This aspect has been emphasized in particular by the culture-historical school of psychology (Vygotsky, 1987; Leontyew, 1981).

[10] See Tomasello and Call (1997: in particular 136-161) on the question whether humans share such knowledge representation structures even with certain primates and Damerow (1998) and Tomasello and Call (1997: 423-427) on the phylogenesis of human cognition from the viewpoint of cognitive psychology.

[11] A typical problem of this kind is, for example, the debate about whether universal structures of language originate from the spreading of a "proto-language" or whether in those structures universal, biologically founded cognitive structures express themselves (Bickerton, 1988; Renfrew, 1988; Bateman et al., 1990). Similar questions arise in the case of the transition to literacy (Damerow, 2006). Can writing systems originate from completely independent roots? Are there, in particular, completely independent inventions of systems of arithmetic symbols?

[12] This and the definitions given in the following do not correspond with the terms introduced by Bruner for the characterization of external representations (Bruner, Olver, & Greenfield, 1966: 1-67). In particular, his classification into enactive, iconic and symbolic representations is not applied here, since, as will be discussed in the following, it appears unsuited to adequately conceptualise the reflective structure of representations and the reflective dynamic of the relationship between symbols and the objects they represent.

[13] That the use of the term *number* can be seen as an indication of a higher level of meta-cognition compared to the simple use of designations of numbers is apparent from the fact that non-literate cultures, even those with developed systems of counting, do not as a rule possess terms of this kind. This is in keeping with the fact that in those cultures even abstract counting without identification of concrete objects of counting is often regarded as meaningless. Even the early high civilizations with developed mathematics, for instance Egypt and Babylonia, do not have a term corresponding to our word *number*.

[14] The use of variables instead of specific designations of numbers offers the possibility to determine precisely the degree of generalization of statements and to change it by substitution of variables in a controlled way. It thus opens a potential means of representation for a higher level of meta-cognitive insights such as the recognition that the natural numbers can be complemented with negative numbers to the system of whole numbers.

[15] A prominent example of such a transfer of the constructive character of a representation to a meta-cognitive level is offered by the emergence of the deductive method in Greek mathematics. At first, arithmetical insights were constructed by figured numbers, that is, patterns of geometrically arranged counters, and geometrical insights by constructions with compass and ruler. The definitions and rules of deductive systems (e.g. those of Euclid's *Elements*) are no longer constructive in this sense, but they are constructive with regard to the structuring of the mathematical knowledge gained in the process.

[16] The meta-constructive character of higher-order representations may, for example, be used to prove, through the construction of models, the relative consistency of a deductive system. The so-called "Klein model" is a Euclidean model of the hyperbolic, non-Euclidean geometry; it demonstrates that a contradiction in hyperbolic geometry would generate a contradiction in the Euclidean geometry. By means of meta-mathematical semiotic reflection, a geometric figure is here constructed which, contrary to Euclid's constructions, serves only meta-cognitive purposes.

[17] According to Piaget, the logico-mathematical operations of operative intelligence pre-exist as practical intelligence in the pre-operative phase of ontogenetic development in the form of senso-motorical schemata of action. These "proto-logic" systems of action have been systematically examined, in particular by Langer (1980, 1986).

[18] Such systems and their cognitive effects have been made the object of research in developmental psychology, in particular by Piaget; see (Piaget, 1950). To the extent that such systems (for instance, the technique of constructing one-to-one correspondences to a standard of counting) are indeed nearly universal (Pinxten, 1976; Dasen, 1977), they can be identified in surveys of the historical

development of numbers (e.g. Menninger, 1970; Tropfke, 1930-1940; Ifrah, 2000) with relative ease.

[19] All arithmetical skills that result from processes of learning are probably based on these culture-specific systems of action and representations (Campbell, 1992). Even though there are precious few cross-cultural studies concerning those skills (Ashcraft, 1992), there is no doubt that they are largely of culture-specific nature and depend on the arithmetical tools that are known to us from ethnological and historical sources.

[20] For example the "Klein model" of hyperbolic geometry mentioned above.

[21] This listing should not be misunderstood as a paraphrase of an axiomatic representation but rather as a hint for identifying basic arithmetical activities which may be considered as candidates for the historical origins of numbers. The structure of natural numbers can be characterized axiomatically in various ways. Through the Peano axioms, numbers are reduced to the iteration of units. Their definition as equivalence classes of sets with equal cardinality characterizes them as a structure of one-to-one correspondences. In a definition as semigroup with specific characteristics natural numbers are characterized by arithmetical operations. Those reductions of the structure of natural numbers to a few basic conditions are of only limited interest for the reconstruction of the origin and development of numbers, for numbers have historically certainly not been deduced from an (axiomatically describable) original definition, but probably rather originated through the integration of structural elements of their overall structure.

[22] The sequence of counting of the Iqwaye, studied by Mimica, was, for example, reconstructed mainly from the information given by two persons (Mimica, 1988: 27).

[23] See e.g. the detailed description of the cultural environment provided by Lancy (1983).

[24] See e.g. the interpretation of notches on a bone excavated at Ishango by Heinzelin (1962), who assumed that the irregularity of the numerical pattern of the notches is an indication of especially sophisticated arithmetical knowledge including, for instance the knowledge of prime numbers. See further, in particular, Marshack's (1972) interpretations of numerous findings. See also the controversy about Marshack's interpretations, which is documented by the criticism of D'Errico (1989a) and the rejoinders of Marshack (1989) and D'Errico (1989b).

[25] The discovery of this function is of recent date. The first archaeologists who interpreted these tokens as arithmetical representations are Amiet (1966) and Schmandt-Besserat (1977). Current literature on the functions of these clay tokens is extensive. Nevertheless, no convincing reconstruction of what objects and quantities they represent has been possible so far. To know what combinations of such tokens are hidden in numerous excavated but still unopened clay bullae might help to solve this problem (Damerow & Meinzer, 1995).

[26] Schmandt-Besserat (1992) provides a comprehensive overview of the excavated tokens and numerical tablets, Englund (1998) of proto-cuneiform tablets. For the less well documented proto-Elamite tablets see Damerow & Englund (1989), Englund (2004), and Dahl (in press).

[27] For the decipherment of proto-cuneiform systems of numerical signs see the contributions of Vaiman (1974), Friberg (1978, 1979), Damerow & Englund (1987), and Nissen, Damerow, & Englund (1993).

[28] Although detailed studies of the sources have been presented, there still exists the occasional prejudice that the Greeks were the first to develop abstract mathematics and that all pre-Greek mathematics was oriented exclusively towards practical purposes. This view can no longer be supported considering the original sources that have come down to us (Høyrup & Damerow, 2001).

[29] System S using the notation introduced by Damerow & Englund (1987); not to be confused with the later sexagesimal place value system.

[30] The recorded Sumerian sequence of number words was strictly sexagesimally structured (Powell, 1973). Since this sequence of number words has only been preserved in very late sources, however, the possibility cannot be excluded that it was artificially created by the scribes of periods following that of the emergence of writing, based on sexagesimal numerical notations in the cuneiform texts available to them.

[31] This date is based on archaeological evidence as well as on the consistency of the historical development of calculation techniques (Nissen, Damerow, & Englund, 1993; Damerow 1995: 199-261; Høyrup & Damerow, 2001: 219-310; for the transition period see Robson, 1999). A different opinion about the dating is held by Powell (1976).

[32] The so-called mathematical cuneiform texts mainly come from this period. Compare for this the standard literature on the history of mathematics, in particular the work of Neugebauer (1935-1937), Neugebauer & Sachs (1945), Friberg (1990), and Robson (1999). The interpretation of these texts has, however, been subjected to a substantial revision in recent years (Høyrup & Damerow, 2001), in particular, due to new translations of the mathematical *termini technici* for arithmetical operations suggested by Høyrup (1990, 2002). The view argued here of Babylonian mathematics resulting from the reflection of culture-specific arithmetical activities at a level of a unified abstract symbol system is based on such new findings.

[33] A recurring topos of mathematical historiography states that pre-Greek mathematics did not know proofs (Becker & Hofman, 1951: 41; Vogel 1958-1959, vol. 2: 84f; Wussing, 1965: 56.; Gericke, 1970: 20, 1984: 71ff). Historians who study pre-Greek mathematics have with good reason repeatedly raised the objection that here a particular form of representing deductive conclusions is being confused with the mental operations themselves and thus that European mathematical tradition, of which this form of representation is characteristic, is being eurocentrically overestimated (Joseph, 1992; Gerdes, 1990: in particular 24ff; for Chinese mathematics: Chemla, 1991; for Babylonian mathematics: Friberg, 1990: in particular 582ff; Høyrup, 1990). In fact, pre-Greek mathematics is distinguished from Greek mathematics not by a lack of deductive reasoning but rather in the way such reasoning is externally represented. Once, however, the steps of such reasoning are externally encoded in written language or symbols, a new type of symbolic activity on a meta-level is induced. Such a development makes Greek mathematics exceptional in comparison to other cultures at its time.

[34] The definitions VII 1 through VII 2, VII 6 through VII 10 and VII 12 (Heath, 1959, vol. 2: 277-286), as well as the theorems IX 21 through IX 34 (Heath, 1959, vol. 2.: 413-420).

[35] Several Greek mathematicians used this definition. The most explicit passage in Plato's work elucidating the underlying idea of number is the dialogue about arithmetic in his *Republic* (Plato, 1994, 524d-526c: 158-167).

[36] In Plato's *Republic* Socrates instructs his interlocutor Glaucon: 'For you are doubtless aware that experts in this study, if anyone attempts to cut up the 'one' in argument, laugh at him and refuse to allow it; (...) Suppose now, Glaucon, someone were to ask them, 'My good friends, what numbers are these you are talking about, in which the one is such as you postulate, each unity equal to every other without the slightest difference and admitting no division into parts?' What do you think would be the answer?" "This, I think—that they are speaking of units which can only be conceived by thought, and which it is not possible to deal with in any other way." (Plato, 1994, 526a: 164-167).

[37] Michael Stifel called the negative numbers "numeri absurdi" (Tropfke, 1930-1940, vol. 2: 98).

REFERENCES

Amiet, P. (1966). Il y a 5000 ans les Elamites inventaient l'Ecriture. *Archaeologia, 12*, 16-23.

Arcà, M. (1984). Strategies for categorizing change in scientific research and in children's thought. *Human Development, 27*, 335-341.

Ascher, M. (1986). Mathematical ideas of the Incas. In M. P. Closs (Ed.), *Native American mathematics*. Austin: University of Texas Press, 261-289.

Ascher, M., & Ascher, R. (1971/1972). Numbers and relations from ancient Andean quipus. *Archive for History of Exact Sciences, 8*, 288-320.

Ashcraft, M. H. (1992). Cognitive arithmetic: A review of data and theory. *Cognition, 44*, 75-106.

Bachelard, G. (1994). *La philosophie du non* (4th ed.). Paris: Presses Univ. de France.

Bachelard, G. (1996). *La formation de l' esprit scientifique*. Paris: Vrin.

Bateman, R., Goddard, I., O'Grady, R., Funk, V. A., Mooi, R., Kress, W. J., et al. (1990). Speaking of forked tongues: The feasibility of reconciling human phylogeny and the history of language. *Current Anthropology, 31*(1), 1-24.

Becker, O. (1936). Die Lehre vom Geraden und Ungeraden im Neunten Buch der Euklidischen Elemente. *Quellen und Studien zur Geschichte der Mathematik, Astronomie und Physik*, Abt. B, Band 3, 533-553.

Becker, O., & Hofman, J. E. (1951). *Geschichte der Mathematik*. Bonn: Athenäum.

Bickerton, D. (1988). A two-stage model of the human language faculty. In S. Strauss (Ed.), *Ontogeny, phylogeny, and historical development*. Norwood: Ablex, 86-105.

Blake, B. J. (1981). *Australian aboriginal languages*. London: Angus & Robertson.

Brainerd, C. J. (1979). *The origins of the number concept*. New York: Praeger.

Bruner, J. S., Olver, R. R., & Greenfield, P. M. (Eds.) (1966). *Studies in cognitive growth*. New York: John Wiley & Sons.

Campbell, J. I. D. (Ed.) (1992). *The nature and origins of mathematical skills*. Amsterdam: North-Holland.

Carnap, R. (1955). Foundations of logic and mathematics. In O. Neurath, R. Carnap & C. Morris (Eds.), *International encyclopedia of unified science* (vol. 1). Chicago: The University of Chicago Press, 139-213.

Cassirer, E. (1953-1957). *The philosophy of symbolic forms*. New Haven: Yale University Press.

Chace, A. B. (1927). *The Rhind mathematical papyrus*. Oberlin: Mathematical Association of America.

Chemla, K. (1991). Theoretical aspects of the Chinese algorithmic tradition (first to third century). *Historia Mathematica, 42*, 75-98.

Closs, M. P. (1986). The mathematical notation of the ancient Maya. In M. P. Closs (Ed.), *Native American mathematics*. Austin: University of Texas Press, 291-369.

D'Errico, F. (1989a). Palaeolithic lunar calendars: A case of wishful thinking? *Current Anthropology, 30*(1), 117-118.

D'Errico, F. (1989b). On wishful thinking and lunar "calendars": Reply. *Current Anthropology, 30*(4), 494-500.

Dahl, J. L. (in press). Animal husbandry in Susa during the proto-Elamite period. *Studi Micenei ed Egeo-Anatolici*.

Damerow, P. (1995). *Abstraction and representation: Essays on the cultural evolution of thinking*. Dordrecht: Kluwer.

Damerow, P. (1996). Number as a second-order concept. *Science in Context, 9*(2), 139-149.

Damerow, P. (1998). Prehistory and cognitive development. In J. Langer & M. Killen (Eds.), *Piaget, evolution, and development*. Mahwah: Erlbaum, 247-269.

Damerow, P. (2000). How can discontinuities in evolution be conceptualized? *Culture & Psychology, 6*(2), 155-160.

Damerow, P. (2006). The origins of writing as a problem of historical epistemology. *Cuneiform Digital Library Journal*, http://cdli.ucla.edu/pubs/cdlj/2006/cdlj2006_001.html.

Damerow, P., & Englund, R. K. (1987). Die Zahlzeichensysteme der Archaischen Texte aus Uruk. In M. W. Green & H. J. Nissen (Eds.), *Zeichenliste der Archaischen Texte aus Uruk* (ATU 2). Berlin: Gebr. Mann, 117-166.

Damerow, P., & Englund, R. K. (1989). *The proto-Elamite texts from Tepe Yahya*. Cambridge: Harvard University Press.

Damerow, P., & Lefèvre, W. (1998). Wissenssysteme im geschichtlichen Wandel. In F. Klix & H. Spada (Eds.), *Enzyklopädie der Psychologie. Themenbereich C: Theorie und Forschung*, Serie II: Kognition, Band 6: Wissen. Göttingen: Hogrefe, 77-113.

Damerow, P., & Meinzer, H.-P. (1995). Computertomografische Untersuchung ungeöffneter archaischer Tonkugeln aus Uruk: W 20987,9, W 20987,11 und W 20987,12. *Baghdader Mitteilungen, 26*, 7-33, Tf. 31-34.

Damerow, P., & Schmidt, S. (2004). Arithmetik im historischen Prozess: Wie "natürlich" sind die "natürlichen Zahlen"? In G. N. Müller, H. Steinbring & E. C. Wittmann (Eds.), *Arithmetik als Prozess*. Seelze: Kallmeyer, 131-182.

Dasen, P. R. (1977). Are cognitive processes universal? A contribution to cross-cultural Piagetian psychology. In N. Warren (Ed.), *Studies in cross-cultural psychology*. London: Academic Press, 155-201.

Dasen, P. R., & de Ribaupierre, A. (1988). Neo-Piagetian theories: Cross-cultural and differential perspectives. In A. Demetriou (Ed.), *The neo-Piagetian theories of cognitive development: Toward an integration*. Amsterdam: North-Holland, 287-326.

Dasen, P. R., & Heron, A. (1981). Cross-cultural tests of Piaget's theory. In H. C. Triandis & A. Heron (Eds.), *Handbook of cross-cultural psychology* (vol. 4). Boston: Allyn and Bacon, 295-341.

Davis, R. B. (1984). *Learning mathematics: The cognitive science approach to mathematics education*. London: Croom Helm.

de Heinzelin, J. (1962). Ishango. *Scientific American, 206* (June), 105-116.

Dixon, R. M. W. (1980). *The languages of Australia*. Cambridge: Cambridge University Press.

Dixon, R. M. W., & Blake, B. J. (Eds.) (1979 ff.). *Handbook of Australian languages*. Amsterdam: Benjamins.

Dux, G. (1992). *Die Zeit in der Geschichte: Ihre Entwicklungslogik vom Mythos zur Weltzeit*. Frankfurt: Suhrkamp.

Dux, G., & Wenzel, U. (Eds.) (1994). *Der Prozeß der Geistesgeschichte: Studien zur ontogenetischen und historischen Entwicklungslogik des Geistes*. Frankfurt: Suhrkamp.

Englund, R. K. (1988). Administrative timekeeping in ancient Mesopotamia. *Journal of the Economic and Social History of the Orient, 31*, 121-185.

Englund, R. K. (1990). *Organisation der Verwaltung der Ur III-Fischerei*. Berlin: Reimer.

Englund, R. K. (1998). Texts from the late Uruk period. In J. Bauer, R. K. Englund & M. Krebernik (Eds.), *Mesopotamien: Späturuk-Zeit und Frühdynastische Zeit*. Göttingen: Vandenhoeck & Ruprecht, 15-233.

Englund, R. K. (2004). The state of decipherment of proto-Elamite. In S. Houston (Ed.), *The first writing: Script invention as history and process*. Cambridge: Cambridge University Press, 100-149.

Ferreira, M. K. L. (1997). When 1+1≠2: Making mathematics in Central Brazil. *American Ethnologist, 24*, 132-147.

Frege, G. (1959). *The foundations of arithmetic: A logico-mathematical enquiry into the concept of number* (2nd ed.). Oxford: Blackwell.

Friberg, J. (1978). *A method for the decipherment, through mathematical and metrological analysis, of proto-Sumerian and proto-Elamite semi-pictographic inscriptions*. Göteborg: Chalmers University of Technology.

Friberg, J. (1979). *Metrological relations in a group of semi-pictographic tablets of the Jemdet Nasr type, probably from Uruk-Warka*. Göteborg: Chalmers University of Technology.

Friberg, J. (1990). Mathematik. In D. O. Edzard (Ed.), *Reallexikon der Assyriologie und Vorderasiatischen Archäologie* (Vol. 7). Berlin: de Gruyter, 531-585.

Fuson, K. C., & Hall, J. W. (1983). The acquisition of early number word meanings: A conceptual analysis and review. In H. P. Ginsburg (Ed.), *The development of mathematical thinking*. New York: Academic Press, 49-107.

Gaida, M., & Tear, D. (1984). Kalender, Numerologie und lunare Astronomie auf Copán-Monumenten. *Beiträge zur allgemeinen und vergleichenden Archäologie, 6*, 310-353.

Gallistel, C. R., & Rochel, G. (1992). Preverbal and verbal counting and computation. *Cognition, 44*, 43-74.

Gay, J., & Cole, M. (1967). *The new mathematics and an old culture*. New York: Holt, Rinehart & Winston.

Gelman, R. (1972). Logical capacity of very young children: Number invariance rules. *Child Development, 43*, 75-90.

Gelman, R., & Gallistel, C. R. (1978). *The child's understanding of number.* Cambridge: Harvard University Press.

Gentner, D., & Stevens, A. L. (Eds.) (1983). *Mental models.* Hillsdale: Erlbaum.

Gerdes, P. (1990). *Ethnogeometrie: Kulturanthropologische Beiträge zur Genese und Didaktik der Geometrie.* Bad Salzdetfurth: Franzbecker.

Gericke, H. (1970). *Geschichte des Zahlbegriffs.* Mannheim: Bibliographisches Institut.

Gericke, H. (1984). *Mathematik in Antike und Orient.* Berlin: Springer.

Gillings, R. J. (1972). *Mathematics in the times of the pharaohs.* Cambridge: MIT Press.

Gnerre, M. C. (1986). Some notes on quantification and numerals in an Amazon language. In M. P. Closs (Ed.), *Native American mathematics.* Austin: University of Texas Press, 71-91.

Gumperz, J. J., & Levinson, S. C. (Eds.) (1996). *Rethinking linguistic relativity.* Cambridge: Cambridge University Press.

Hallpike, C. R. (1979). *The foundations of primitive thought.* Oxford: Clarendon Press.

Heath, S. T. L. (1956). *The thirteen books of Euclid's Elements* (2nd ed.). New York: Dover.

Heath, S. T. L. (1921). *A history of Greek mathematics.* Oxford: Dover.

Høyrup, J. (1990). Algebra and naive geometry. *Altorientalische Forschungen, 17,* 27-69, 262-354.

Høyrup, J. (2002). *Length, width, surfaces: A portrait of old Babylonian mathematics and its kins.* New York: Springer.

Høyrup, J., & Damerow, P. (Eds.) (2001). *Changing views on ancient Near Eastern mathematics.* Berlin: Reimer.

Hurford, J. R. (1975). *The linguistic theory of numerals.* Cambridge: Cambridge University Press.

Hurford, J. R. (1987). *Language and number.* Oxford: Basil Blackwell.

Ifrah, G. (2000). *The universal history of numbers: From prehistory to the invention of the computer.* New York: John Wiley & Sons.

Joseph, G. G. (1992). *The crest of the peacock: The non-European roots of mathematics.* London: Penguin.

Juschkewitsch, A. P. (1966). *Geschichte der Mathematik im Mittelalter.* Basel: Pfalz.

Kant, I. (1985). Prolegomena to any future metaphysics that will be able to come forward as science. In I. Kant, *The philosophy of material nature.* Indianapolis: Hackett, 1-121.

Kant, I. (1998). *Critique of pure reason.* Cambridge: Cambridge University Press.

Lancy, D. F. (1983). *Cross-cultural studies in cognition and mathematics.* New York: Academic Press.

Langer, J. (1980). *The origins of logic: Six to twelve months.* New York: Academic Press.

Langer, J. (1986). *The origins of logic: One to two years.* New York: Academic Press.

Lefèvre, W. (1981). Rechensteine und Sprache. In P. Damerow & W. Lefèvre (Eds.), *Rechenstein, Experiment, Sprache: Historische Fallstudien zur Entstehung der exakten Wissenschaften.* Stuttgart: Klett-Cotta, 115-168.

Leontyew, A. N. (1981). *Problems of the development of the mind.* Moscow: Progress Publishers.

Lévy-Bruhl, L. (1926). *How natives think.* New York: MacMillan.

Li Yan & Dù Shíràn (1987). *Chinese mathematics: A concise history.* Oxford: Clarendon.

Locke, L. L. (1923). *The ancient quipu or Peruvian knot record.* New York: The American Museum of Natural History.

Lorenzen, P. (1955). *Einführung in die operative Logik und Mathematik.* Berlin: Springer.

Marshack, A. (1972). *The roots of civilization: The cognitive beginnings of man's first art, symbol and notation.* London: Weidenfeld and Nicolson.

Marshack, A. (1989). On wishful thinking and lunar "calendars". *Current Anthropology, 30*(4), 491-494.

Menninger, K. (1970). *Number words and number symbols: A cultural history of numbers.* Cambridge: MIT Press.

Mimica, J. (1988). *Intimations of infinity: The mythopoeia of the Iqwaye counting system and number.* Oxford: Berg.

Minsky, M. L. (1975). A framework for representing knowledge. In P. H. Winston (Ed.), *The psychology of computer vision.* New York: McGraw-Hill, 211-277.

Minsky, M. L. (1985). *The society of mind*. New York: Simon & Schuster.

Needham, J. (1959). *Mathematics and the sciences of the heavens and the earth*. Cambridge: Cambridge University Press.

Neugebauer, O. (1926). *Die Grundlagen der ägyptischen Bruchrechnung*. Berlin: Springer.

Neugebauer, O. (1934). *Vorgriechische Mathematik*. Berlin: Springer.

Neugebauer, O. (1935-1937). *Mathematische Keilschrifttexte*. Berlin: Springer.

Neugebauer, O., & Sachs, A. (1945). *Mathematical cuneiform texts*. New Haven: American Oriental Society.

Nissen, H. J., Damerow, P., & Englund, R. K. (1993). *Archaic bookkeeping: Early writing and techniques of economic administration in the ancient Near East*. Chicago: Chicago University Press.

Piaget, J. (1950). *Introduction à l'épistémologie génétique*. Paris: Presses Univ. de France.

Piaget, J. (1971). *Biology and knowledge: An essay on the relations between organic regulations and cognitive processes*. Chicago: University of Chicago Press.

Piaget, J., & Garcia, R. (1989). *Psychogenesis and the history of science*. New York: Columbia University Press.

Piaget, J. (1952). *The child's conception of number*. London: Routledge and Paul.

Pinxten, R. (Ed.) (1976). *Universalism versus relativism in language and thought*. Den Haag: Mouton.

Platon (1994). *Republic*. Cambridge: Harvard University Press.

Powell, M. A. (1973). *Sumerian numeration and metrology*. Ann Arbor: University Microfilms.

Powell, M. A. (1976). The antecedents of Old Babylonian place notation and the early history of Babylonian mathematics. *Historia Mathematica, 3*, 417-439.

Renfrew, C. (1988). Archaeology and language. *Current Anthropology, 29*(3), 437-441.

Robson, E. (1999). *Mesopotamian mathematics, 2100-1600 BC: Technical constants in bureaucracy and education*. Oxford: Oxford University Press.

Saxe, G. B. (1982). Culture and the development of numerical cognition: Studies among the Oksapmin of Papua New Guinea. In C. J. Brainerd (Ed.), *The development of logical and mathematical cognition*. New York: Springer, 157-176.

Scharlau, B., & Münzel, M. (1986). *Quellqay: Mündliche Kultur und Schrifttradition bei Indianern Lateinamerikas*. Frankfurt: Campus.

Schmandt-Besserat, D. (1977). An archaic recording system and the origin of writing. *Syro-Mesopotamian Studies, 1*, 31-70.

Schmandt-Besserat, D. (1992). *Before writing*. Austin: University of Texas Press.

Smith, D. A., Greeno, J. G., & Vitolo, T. M. (1989). A model of competence for counting. *Cognitive Science, 13*, 183-211.

Strauss, S. (Ed.) (1988). *Ontogeny, phylogeny, and historical development*. Norwood: Ablex.

Szabó, Á. (1960). Anfänge des Euklidischen Axiomensystems. *Archive for History of Exact Sciences, 1*, 38-106.

Szabó, Á. (1978). *The beginnings of Greek mathematics*. Dordrecht: Reidel.

Szabó, Á. (1994). *Die Entfaltung der Griechischen Mathematik*. Mannheim: Bibliografisches Institut.

Thompson, J. E. S. (1941). *Maya arithmetic*. Washington: Carnegie Institute.

Thompson, J. E. S. (1960). *Maya hieroglyphic writing*. Norman: University of Oklahoma Press.

Tomasello, M., & Call, J. (1997). *Primate cognition*. New York: Oxford University Press.

Tropfke, J. (1930-1940). *Geschichte der Elementar-Mathematik* (3rd ed.). Berlin: de Gruyter.

Tropfke, J., Vogel, K., Reich, K., & Gericke, H. (1980). *Geschichte der Elementar-Mathematik* (4th ed.). Berlin: de Gruyter.

Vaiman, A. (1974). Über die Protosumerische Schrift. *Acta Antiqua Hung., 27*, 15-27.

van der Waerden, B. L. (1947/1949). Die Arithmetik der Pythagoreer. *Mathematische Annalen, 120*, 127-153, 676-700.

van der Waerden, B. L. (1954). *Science awakening*. Groningen: Noordhoff.

Vogel, K. (1958/1959). *Vorgriechische Mathematik*. Hannover/Paderborn: Schroedel/Schöningh.

Vygotsky, L. S. (1987). Thinking and speech. In R. W. Rieber & A. S. Carton (Eds.), *The collected works of L. S. Vygotsky* (vol. 1). New York: Plenum, 37-285.

Wertheimer, M. (1925). Über das Denken der Naturvölker: Zahlen und Zahlgebilde. In M. Wertheimer, *Drei Abhandlungen zur Gestalttheorie.* Erlangen: Palm & Enke, 106-163.

Whorf, B. L. (1956). *Language, thought and reality: Selected writings of Benjamin Lee Whorf* (J. B. Carroll ed.). Cambridge: MIT Press.

Wussing, H. (1965). *Mathematik in der Antike.* Leipzig: Teubner.

AFFILIATIONS

Peter Damerow
Max Planck Institute for the History of Science,
Berlin

YVES CHEVALLARD

IMPLICIT MATHEMATICS

Their Impact on Societal Needs and Demands

My presentation will center on a few, bold contentions, that I shall try to substantiate concisely. To begin with, let me say that, in essence, I shall dispute the validity of a wide-spread belief, forcefully expressed in the following quotation borrowed from ICME V's report on mathematical modelling[38]:

> The ultimate reason for teaching mathematics to students, at all educational levels, is that mathematics is useful in practical and scientific enterprises in society.

The following considerations aim to reveal and, to a certain extent, to explain the essential ambiguity in this and similar declarations.

1. On the alleged utility of mathematics

My main contention in this regard can be expressed tersely.

1.1. No modern society can live *without* mathematics.

1.2. In contradistinction to societies as organized bodies, all but a few of their members can and do live a gentle, contented life *without any mathematics whatsoever.*

Certainly both theses require careful explanation. They seem to refer to the "degree of presence" of mathematics in society. But to gain insight, one must resort to another, distinct notion, that of the *mode of presence* of mathematics: mathematics may be present either in explicit form or in implicit form.

2. Explicit (uses of) mathematics

Explicit, or *live*, or *visible* mathematics, or more precisely their *explicit mode of presence* is what people have in mind when they praise mathematics for being a necessity of life today.

2.1. Explicit mathematics are the mathematics that are visibly handled, used, manipulated in science (including mathematics), technology, engineering, business, administration.

2.2. Explicit uses of mathematics are essential to our present-day societies, but: 1. they are generally concealed from public view; 2. in going about their business,

U. Gellert, E. Jablonka (eds.), Mathematisation – Demathematisation: Social, Philosophical and Educational Ramifications, 57–65. © 2007 Sense Publishers. All rights reserved.

most people never meet with explicit uses of mathematics – save for some arithmetic[39].

2.3. Accordingly, not only does the word "mathematics" mean, for most people, explicit mathematics: it is usually reduced to apply to the only explicit mathematics that become normally visible to the layman, i.e. school mathematics[40].

2.4. Some mathematics educators claim – without much direct evidence – that "mathematics" permeate every aspect of life. As a further consequence, in so far as they want school mathematics to reflect faithfully the mathematics of the "outer world" as they see it, they become inveigled into drawing an image of reality stuffed with explicit mathematics, in order that reality may conform to the fiction they have created. Hence the plight of those who devote their energies to proving that mathematics, especially in the form of mathematical modelling, are "at work" in every nook and cranny of society – a Sisyphean labour and, up to a point, a wild-goose chase.

3. The social "implicitation" of mathematics into objects

If it is true that mathematics pervade present-day, Western-type societies, it is nevertheless true in a very different sense. The way in which mathematics penetrate our daily life is unremarkable, even banal. Mathematics in effect find their shortest way to every one of us through "objects" of all kinds, in the form of implicit mathematics.

3.1. Implicit mathematics are formerly explicit mathematics that have become "embodied", "crystallized" or "frozen" in objects of all kinds – mathematical and non-mathematical, material and non-material –, for the production of which they have been used and "consumed".

3.2. The amount of implicit mathematics present in an object, i.e. the amount of mathematics crystallized in it, can be roughly defined as the sum total of the explicit mathematics used in producing that object, and a fraction of the mathematics (previously) crystallized in the (material and non-material) "objects" consumed in the production of that same object[41].

3.3. Accordingly, beyond any object, however commonplace, one can invoke, in an infinite regress, all those fragments of mathematical knowledge and know-how which have been passed on from object to object, implicitly and most often invisibly, in the social production of, firstly, the object itself, secondly, the objects consumed in producing that object, thirdly, the objects consumed in producing the objects consumed to produce that object, and so forth[42].

3.4. The amount of crystallized mathematics present in a given object is exactly what I call the mathematical *grade* (or *content*, or *tenor*) of the object.

4. Mathematics are (implicitly) everywhere around us

The implicit mode of presence of mathematics usually goes unnoticed; and such is the state and status of "mathematics" in the life of most people.

4.1. The amount of *explicit* mathematics used in producing an object *is almost always negligible.*

4.2. The amount of mathematics recursively crystallized in most objects – including the "necessities of life" – *is almost always considerable*: however negligible may be the amount of explicit mathematics used at any point in their social production process, it can result in the end in a considerable amount of mathematics crystallized in the "finished product"[43].

4.3. The degree of presence of a given means of production is in no way an intrinsic characteristic of a given type of object (as defined by its function and structure), but is actually *socially determined,* i.e. determined by the normal process of production prevalent at the time.

4.4. Although for example men of the Neolithic Age made stone implements – such as hammers and axes – which contained no mathematics at all[44], exactly the same objects with regard to both structure and function are now "stuffed" with crystallized mathematics. Such is the very reason why mathematics can be said to be everywhere around us, though in the unassuming form of dead, frozen mathematics, hidden in the multitudinous objects of everyday life.

5. The growing, invisible, silent presence of mathematics

While the live mathematics incorporated in a few social practices are often called on to speak, crystallized mathematics tell no tales: they are nonetheless an essential component of almost all things and situations that make up contemporary life.

5.1. The previous, much too concise account of the situation of mathematics in society is by no means peculiar to mathematics: considering the overall *production of society*, one can equally sensibly apply it to any material or non-material "means of production" of society, be it medicine, electricity, or steel, for instance.

5.2. To a certain extent, however, one can assess the growing importance and "utility" of mathematics in modern societies by a simple mental experiment: just as we could switch off electricity to satisfy ourselves that electricity is indeed a basic ingredient of developed societies, without which almost nothing would continue to exist, so also we can imagine that the "switching off" of mathematics would cause almost every socially produced thing to fail to exist – a fact coextensive with modern societies.

5.3. The degree of presence of implicit mathematics in society, i.e. the average mathematical grade or tenor of goods and services made available to the man in the street, has varied over time: a pervasive, centuries-old trend, linked to the development and formidable growth of science and technology has for good or bad resulted in a *continuing rise in the mathematical grade* of objects. While for instance the content in steel of many necessities has dramatically fallen – think of motor-cars –, the empire of mathematics is steadily spreading and keeps encroaching on domains which until recently had remained foreign to its influence.

6. The dialectic between implicit and explicit mathematics

Social uses of mathematics lead to a paradoxical situation, which is more often ignored than analysed: one can say that, while society as a machinery is more and more mathematised, our daily life is more and more demathematised.

6.1. The greatest achievement of mathematics, one which is immediately geared to their intrinsic progress, can paradoxically be seen in the never-ending, two-fold process of (explicit) *demathematising* of social *practices* and (implicit) *mathematising* of socially produced *objects* and techniques.

6.2. This applies equally properly to *mathematical* practices and *mathematical* objects: whereas, for example, multiplication was held in Ancient Egypt to be a scholarly technique requiring much skill and intelligence, it has over time become so simplified – so "demathematised" – that even young children can now perform it – a fact so familiar to us that we usually do not question its meaning and significance.

6.3. The process of mathematisation/demathematisation is in fact the very foundation on which the social production of mathematical objects rests: while the mathematical *grade* of mathematical tools steadily *increases*, their mathematical value – that is, the average, socially determined mathematical labour-time needed to produce them – steadily *decreases*.

6.4. The process of demathematisation relates, of course, to the amount of explicit mathematics, i.e. of mathematical knowledge and know-how, needed to produce or to *use*[45] mathematical objects. As regards implicit mathematics, more and more objects tend to have a higher mathematical grade, thus becoming more and more *mathematically powerful*.

6.5. Both the rise in mathematical grade and the decrease in mathematical value must be invoked to explain the *social success of mathematics*: high mathematical grades make for the powerful and efficiency of the objects made available to us, be they material or non-material; and lower and lower mathematical values account for their *wide social availability*.

7. The individual dispensability of mathematics

I shall now take up the question why so few people are directly concerned with (explicit, live) mathematics; and why the ordinary citizen has in fact to deal with so little mathematics in his/her ordinary experience of society.

7.1. One cannot overstress the fact that, in achieving simplification, mathematicians and "mathematics workers" have constantly resorted to one single method. This method of "simplifying" mathematical objects consists in incorporating (explicit) mathematics into them, i.e. in turning live mathematics into dead, crystallized mathematics: any ignoramus can now do any calculation whatever with his pocket calculator, and one should no longer worry even about the niceties of the addition of fractions (a state of things that, for the time being, many mathematics educators still resist).

7.2. The explicit and implicit mathematics embodied into any theorem or method (and which account for their increased powerfulness) are those mathematics needed to establish that theorem or method. But little or no knowledge of *those mathematics* is required in order to use relevantly the (mathematical) tool thus provided: one may use, as a mathematical tool, the theorem of Pythagoras without having any idea of any one of its various potential proofs. Obviously, it is a regular outcome of the activity of mathematicians, throughout the centuries, that formerly difficult questions become easy or easier ones, and that mathematical tools which were at first the privilege of experts sooner or later become available to novices.

7.3. More generally, the increase in mathematical grade and decrease in mathematical value, and the ensuing increase in mathematical powerfulness and social availability of mathematical object, are coexistent with yet another, socially elemental characteristic: while the average mathematical grade of goods and services increases, *the average mathematical expertise required to consume those goods and services steadily decreases*.

7.4. This continuing line of historical development in the production of society explains why the average citizen simply does not have to care much – or very much – about mathematics, while mathematics are (implicitly) all around us in everyday life: most of the ordinary social practices in which he or she happens to take part have been deeply demathematised, a continuing process which has even accelerated since the advent of the microcomputer – thanks to which so many objects with high mathematical grade (and even with high mathematical added value) are made accessible to the "multitude".

8. The cultural fragility of mathematics

Those characteristics which explain the – more or less invisible – social success of mathematics also make for their *cultural fragility*.

8.1. Because they are generally concealed from public view, mathematics are scarcely given credit for what we owe them throughout our daily life. The social *effectiveness* of mathematics is essentially co-terminous with their social – and therefore cultural – *invisibility*.

8.2. The social debate on mathematics thus tends to center on *explicit* mathematics, and obscures the true rôle and major mode of presence of mathematics in society.

8.3. Moreover, the only explicit mathematics that most people ever come close to – in contradistinction to, e.g., the case of electricity – are *school mathematics*, i.e. mathematics as a subject-matter to be taught and learnt. That all but a few people experience explicit mathematics only under these conditions is a fact worthy of note, and the source of many societal problems for which appropriate solutions are yet wanting.

8.4. Indeed, the *teaching of mathematics* to the many is the way Western-type societies have tried to make mathematics *culturally visible*. The historical establishing of mathematics teaching could in fact be expected both to provide society with the necessary, mathematically skilled labour, and to achieve, on behalf of mathematics, cultural recognition and legitimacy. (Because of their social invisibility, mathematics could not manage to survive socially without this recognition.) The venture, one must admit, has not been a complete success.

9. The case of the teaching of mathematics

In trying to reconcile society with mathematics, the central question to be answered, from which so many consequences flow, is: why have mathematics been obstinately taught at the secondary level (as opposed to the primary and tertiary levels), which is undoubtedly the weak link in our educational systems?

9.1. Certainly the growing empire of science over the life of Western societies, from the seventeenth century onwards, and the ensuing need for ever more engineers both civil and military, do account for the fact that *something* had to be done. But many examples – e.g., that of medicine – show that proper training of the required élites could have been started at the university level. That another, distinct line of action was decided on suggests that the main problem attacked was actually of a very different kind.

9.2. Because our societies need mathematics, because they are, so to speak, driven by mathematics, a balance had to be reached, in the sphere of culture, between society and mathematics. Society had, in some way or another, to recognize mathematics as a basic, major ingredient and driving force of economic and social development. By inculcating some mathematics in its children, society thus *paid a tribute to its needs* – to the increasing (implicit) mathematisation of society –, and one can reasonably doubt whether it will ever be out of debt in this respect.

9.3. To some degree, the recognition granted to mathematics proved misleading. For reasons still to be elucidated, the teaching of mathematics came to be justified in terms of the so-called "utility" of mathematics, and this in turn was understood *in terms of the individual's interests* – whereas the overall, *communal interests of society* as a global village were really at stake.

9.4. This determined, if I dare say so, a *cultural pathology* which not only misinterprets social needs of paramount importance, but may also take its toll on the pupil. The real nature of the problem facing us gradually faded from sight; accordingly, the solution afforded by the teaching of mathematics partly lost its intended efficiency.

10. How it all came about

The most essential problem that lastingly confronts the teaching of mathematics (and, more generally, the teaching of any subject-matter whatsoever) is the question of *its very existence as a social practice*, that is, the "socio-ontological" question.

10.1. Historically, people had to fight hard for the birth of mathematics teaching, and still have to attend to maintenance problems both corrective and preventive.

One of the main assignments in this respect is to convince society as a whole that it *needs* mathematics teaching; or rather, that the teaching of mathematics is both necessary and most desirable.

10.2. In seeking to convince society of this vital idea, the pressure group that I call the *noosphere*, i.e. the people who devote time and energy to thinking about the teaching of mathematics, its present state and its foreseeable future, will try to impose simple views and, to this end, will propagate what I call an *apologetical discourse*. It should be emphasized that the apologetics of the noosphere has a very narrow thematic which, moreover, seems quite independent of the subject-matter on behalf of which it is proclaimed. In Western and Western-type societies at least, its recurrent, central claim is that the subject-matter in question should be taught and learnt, because both society as a whole and its members as individuals need to master it *in order to succeed* – success being appreciated according to changing criteria.

10.3. The balance between society's and the individual's reported needs may be achieved in many different ways. But in most cases modern societies have come to be infused with the peculiar spirit of what I shall term *individualistic democracy*, in which nothing can be entirely good for a given society unless it is presumed to be good also for any one of its members. In such a context, those communal needs will tend to be ignored, and effectively neglected, which cannot be made to appear at the same time as common, personal needs – *as needs of the individual as such*. Hence the argument, so often resorted to by noospheric apologetics, according to which mathematics are useful *to almost everyone in almost all situations*.

10.4. Such an astounding privilege has been lavishly bestowed upon almost every subject-matter ever considered for teaching. It is part and parcel of the standard apologetical discourse that noospherians generally rely on. As often as not however, this kind of description wanders from reality; but, as I have tried to show, it is never more unrealistic and, if I may say so, ivory-towerish, than in the case of mathematics.

11. What's wrong with the current apqiogetical discourse

The quotation given earlier is in fact typical of statements proffered by the noosphere. The so-called "ultimate reason" offered here as an argument for teaching mathematics – their supposed usefulness "in practical and scientific enterprises in society" – is typically a noospheric reason, a reason which mingles two permanent and closely-related distinctive features: the impregnation of society with mathematics as a means of production, and the average citizen's personal relationship to mathematics as a body of knowledge.

11.1. Pronouncements in the noosphere, in fact, usually testify to the existence of some generally adverse set of conditions, of some problem with which the teaching system is confronted (or is likely to be confronted in the near future), and which it is the noosphere's duty to come to grips with.

11.2. At the same time however, noospherians usually – and interestedly – miss the point in such polemical declarations. They often cheerfully dismiss reality as it is and indulge in the fallacies of false consciousness. In other words, in voicing such declarations in defence of the teaching system, they make partially irrelevant strategic moves, whose side effects are generally unexpected.

11.3. As a counter-example to the usual noospheric argument, one might consider the case of medicine: medicine pervades our daily, "practical" life as well as major "scientific enterprises" – as the existence of industrial and forensic medicine, and the fresh growth of space and nuclear medicine show. For all that, it is not true that medicine is taught "at all educational levels".

11.4. More generally, should the utility of a given subject-matter be taken as a "reason" for teaching it "at all educational levels", such a reason would remain, obviously, an altogether insufficient one: considered as a subjective motive, it seems unconvincing, too weak in itself to act as a compelling force (most modern societies do not teach medicine at the primary and secondary levels, although medicine is held to be of paramount importance in most human activities); regarded as an objective cause, sufficient in itself to explain the historical establishing of the teaching of mathematics, and keeping the case of medicine in mind, one may wonder why, in this particular instance, like causes do not produce like effects. The social utility of a subject-matter is neither the ultimate reason for, nor the efficient cause of, its being taught. This conclusion, in my view, applies to mathematics as to other bodies of knowledge.

12. Starting all over again

It is up to us, I believe, to reconsider both the problem and the solution. It is my opinion also that in some sense, the noosphere will have to start all over again. And it will have to start at the beginning.

12.1. In choosing to fall back on that besieged territory – mathematics at school –, in pretending that it can serve as an appropriate base of operations from which mathematics could recover cultural visibility and achieve societal legitimacy, in arguing in the face of facts for the dubious utility of mathematics, the noosphere lacks either lucidity or courage – perhaps both.

12.2. I shall maintain that the teaching of mathematics at the secondary level is nothing but a means – for which, of course, we have to pay a high price – to reconcile culturally society with mathematics regarded as an inescapable societal need. Our societies may come to accept the idea that, in studying mathematics (as well as national history, the official language of one's country, etc.), everyone of us pays his personal contribution to the community.

12.3. The teaching of mathematics might then take on a new turn. It could keep closer to the true social rôle of mathematics and be made to play a more relevant part, that of *representing to the rising generation the way in which explicit mathematics are consumed in the production of society* – including the production of that essential component of society, mathematics. Like many other subject-matters taught at school, mathematics education at the primary and secondary levels should be relevantly defined as a *cultural initiation* – one which might enable all members of society to be in tune with the society to which they belong, to understand its most essential workings, and, as the case may be, to take an active part in its scientific and technological development.

12.4. Such an initiation should result in an awareness of society as a complex whole made up of many deeply-interrelated components, most of them hardly visible and understandable from the outside. It should avoid some major pitfalls, address elemental, not necessarily elementary, questions, and beware of unrealistic realism. If it could cast off the sanctified fallacies that I earlier criticised, the "mathematics-at-work" movement might in this respect show us the way.

NOTES

[38] Proceedings of ICME 5, Theme Group 6: Applications and Modelling, p. 199.

[39] Gamblers use some combinatorics: but is not gambling a trade in its own right?

[40] In the mathematicians' sphere, mathematics are used in order to produce more mathematics; in the engineering sphere, mathematics are used to produce more knowledge and know-how of a different kind (in electronics for instance). In the case of school mathematics, mathematics are neither used nor produced: they are taught and learnt.

[41] The reader who, on reading this statement, is reminded even dimply of Marx's labour theory of value is sure not to have missed the point!

[42] This, as one may note, sounds like a typically recursive definition. The Marxist flavour is again very insistent: see for instance Morishima (1973).

[43] That tiny fragments of mathematics can make up a significant whole is a fact that mathematically minded people should feel at ease about.

[44] With the exception of some religious monuments, in which traces of mathematics have been found, notably the supposed use of Pythagorean triples: see van der Waerden (1983).

[45] Possibly for the production of new mathematical objects.

REFERENCES

Morishima, M. (1973). *Marx's labour theory of value.* Cambridge: Cambridge University Press.
van der Waerden, B. L. (1983). *Geometry and algebra in ancient civilizations.* Berlin: Springer.

AFFILIATIONS

Yves Chevallard
L'Institut Universitaire de Formation des Maîtres,
Aix-Marseille

ROLAND FISCHER

TECHNOLOGY, MATHEMATICS AND CONSCIOUSNESS OF SOCIETY[46]

My topic is the role of mathematics in holding together society including a consideration of its relation to what I call consciousness of society. Technology is concerned with this issue since it is of special importance in holding together today's society and, additionally, it has always been closely related with mathematics. "New technologies" do not play a special role, but they are only the most recent manifestations of a specific phenomenon.

I will not deal with any direct consequences these issues may have for mathematics education. I think that we should first acquire a principal common understanding about the situation.

RULE-ORIENTED SOCIETY

Two social scientists who worked in a project of the IFF (Institute of Inter-disciplinary Studies of Austrian Universities, today Faculty of Interdisciplinary Studies of the University of Klagenfurt), Arno Bammé and Peter Fleissner, wrote an article the title of which starts with the demanding question: "What holds the world together?" (Bammé & Fleissner, 1994). They propose two answers.

The first one: the *economy* holds the world together. Economic activity, the exchange of commodities and services, and above all the increasingly fast flow of money connect us. The world-wide market is the most dominant force for the integration of world society. Even if we have nothing else to do with a country far away on another continent, there still exists the possibility for economic relations, at least insofar as there is a possible area for a future market.

The second answer is the following: Technology is the connecting element. This includes, firstly, communication technologies such as television, telephone or the internet, which make possible the increasingly rapid transfer of ever more information; secondly, the means of transportation of commodities and humans; and finally, technologies of production: they generate a necessity for connection and compatibility to use the resources of the whole world for production. Additionally they cause the necessity of a unified demand such that there is a sufficient number of consumers for mass production. In a certain sense, according to A. Bammé, through the dominance of technology, the economy becomes less important. The EAN-Code, the "European Article-Numbers," makes it possible to observe the flow of single commodities from the producer to the consumer, whereby new modes of steering this process become possible, for example "just-in-

U. Gellert, E. Jablonka (eds.), *Mathematisation – Demathematisation: Social, Philosophical and Educational Ramifications*, 67–80. © 2007 Sense Publishers. All rights reserved.

time-production". This demands at least a new definition of trade (see Bammé, 1993).

Comparing these two proposals for explaining what holds the world together, one can see that they have something in common, at least compared with cohesive forces which functioned in earlier times, such as religions or dynasties: in modern times, a rule-oriented mechanistic mode has become more important. It is no longer necessary for people to engage in the holding together and be responsible for the care of the whole of society; a mechanism does the job. Take, for example, the mechanism of the market: The participants have to care for their benefit, especially their profits; the market integrates; an "invisible hand" governs the market, even to the benefit of all participants, as a (well-known) theory would have it. Similar hopes and promises exist with respect to technologies: The totally implemented Internet will generate a social whole, even new modes of establishing collective will. The new whole "emerges" without having been anticipated by human thought. We only have to obey the rules. Bammé uses the term "technological civilisation", following the philosopher Hülsmann (Bammé, Baumgartner, Berger, & Kotzmann, 1987).

One can observe that even in the area of politics, where one would expect conscious decision making, there is an increasing trend towards "technology" (in a more general sense): Agreements about world trade, the introduction of a common currency, sophisticated contracts for military defence or arms control in order to secure peace are complex rule mechanisms. These mechanisms are the result of a difficult negotiating process, where decisions have to be made, but the intention is that after implementation they will work on their own, governing as an "invisible hand" without the necessity of making decisions in each special case. The increasing importance of political bureaucracies, not only on a national level, emphasises the fact that technology in a broad sense plays a significant role in the integration of the society.

A final support for the thesis of technological integration of society (which in this general form includes the thesis of integration by the market) comes from the system-theoretic conception of the society by Niklas Luhmann: society as a conglomeration of autonomous systems such as the economy, the judicial system, science, education etc. These systems have an internal communication according to specific codes and fulfil certain functions for each other. But no system, not even the political system, is in the position to govern the "whole". Society "functions", it runs and works if all the (sub-) systems do their jobs (see Luhmann, 1986).

UNCONSCIOUSNESS OF THE SOCIETY

One benefit of technological integration of the society is the *relief from strain*. The integration takes place automatically. A further advantage: Personal domination is not necessary to the same extent as it is in other models of integration. Instead of personal domination the rule "dominates": the market, money, bureaucracy, the computer (-network). Thereby *more democracy* is possible. Additionally, the formal character of rules makes possible a *diversity of contents* (preferences of people, ideas, culture etc.)

On the other hand there also exist disadvantages. It is difficult to assign responsibility. This concerns especially situations in which the rule system does not solve important problems, such as the problem of damage to the natural environment or the exclusion of people from social processes (unemployment). Or still more serious: What is to be done if the rule system directly causes certain problems or prevents society from recognizing these problems? (According to N. Luhmann, the problems do not exist as social problems if they remain unrecognised – see Luhmann, 1986).

For me the rule-oriented, technological holding together of society has to do with unconsciousness of society. Society delivers itself up to certain mechanisms, without a common conscious idea about its whole, and, as a consequence, without any common responsibility for it. This unconsciousness makes it difficult or even impossible to make common decisions or take action in situations where the mechanisms fail.

Unconsciousness of society is not the same as the individual unconsciousness of people in society. Many people today have ideas about themselves and also about the larger whole of which they are a part. They have, so to say, "social consciousness". But the sum of these individual "consciousnesses" fails to amount to a consciousness of society. On the other hand a consciousness of society does not mean that all people – or at least the majority of them or the most influential – must have the same ideas about the whole. What I mean by "consciousness of society" is more complicated and I will present the concept below. But before doing so I want to say a few words about the role of science and especially of mathematics in a rule-oriented society.

KNOWLEDGE SOCIETY

For a rule-oriented society *objective knowledge* and science are of special importance. They offer the basis which allows for the constructing of rules and for additional expertise if the rule system is not sufficient for decision making. There exists the desire that this basis be anchored outside ourselves and our arbitrariness as objective knowledge, preferably about "nature" (in a general sense of the word, including, for example, knowledge about the nature of human beings as the basis for economic rules). The compulsion imposed by objects would offer a rule-oriented society the support from the outside which in former times was offered by authorities on a transcendental basis. In contrast to such earlier supports from the outside, the new one is compatible with democracy: Through education everybody can gain insight into the necessities. At best the knowledge of the necessities generates a "common consciousness" and holds the world together.

Today we have begun to recognise that objective knowledge does not really fulfil the task which has been assigned to it. Knowledge is incomplete, experts contradict each other, paradigms change. Above all the constructive character of knowledge – namely that it is influenced by us, our interests, modes of recognising, thinking and communicating – restricts its function as an objective grip from outside (of society). We are thrown back onto ourselves (see Fischer, 1998).

One way out of this situation that would make possible the establishment of connections and commonality, despite all the restrictions of knowledge and science, seems to be through abstraction and methodisation. Though there are deficits, contradictions, changes and uncontrollable conditions in science, there nevertheless exist common principles and methods. Otte claims that there began in the 19th century a process of theoretisation and methodisation of knowledge (Otte, 1993: 131, 149). Modern sciences formulate hypotheses, develop models and theories. But science refuses to make direct statements about reality. Even if our thoughts about the reality differ, we, as scientists, have in common several principles and methods.

At this point mathematics comes into play. It goes most consequently the way of abstraction: concentration on methods and restriction to what (almost) all humans agree on. Moreover, the objects of mathematics are – among others – rules. Mathematics develops systems of rules, explores them logically – in pure mathematics – and then applies them. Mathematics is the science of rule-oriented technology in the most general sense.

Theorising and methodising of knowledge, with mathematics at the forefront of this movement – does it offer a grip from outside? One can view theorising and methodising simply as movements away from objects towards the human; toward how he/she views and orders things, generates structures etc. Where can a grip from outside be found here?

MATHEMATICS AS MATERIALISATION OF ABSTRACTS

The thesis is: *For fostering the process of abstraction and methodisation mathematics provides a grip from outside by materialisation.* The entity about the existence of which we have the highest commonly agreed certainty – outside ourselves – is matter. And matter is used by mathematics to make abstracts graspable and manipulable; by this means mathematics offers a grip from outside. This is true in present times, but has been the case even in former times (when computers did not exist). To put it another way: Hardware has always been significant for mathematics.

What are the supports for the thesis of materialisation? They begin with stones and fingers for supporting counting, continue with abacuses and graphical representations on paper and include the most recent manifestation, the computer. In our culture the graphical materialisations are most significant: simple dash-symbols for counting, the various modes to write numbers including the decimal notation, the formulas of elementary algebra, the symbols of calculus, including the graph of a function, the symbols of linear algebra, set theory and finally the notations to support and fix logical argumentation and proofs. In the 20th century a series of further representation tools arose: graphs with edges and vertices to visualise networks, flow diagrams in order to represent processes, new graphical methods in statistics, to name a few. There are greater and lesser degrees of "precise" materialisations: Symbols for numbers are usually meant to be precise, but to draw the neighbourhood of a point on the number line in order to support argumentation is not so precise. The "geometrisations" of many areas of mathematics, analysis, linear algebra, stochastics etc. – in all these fields one speaks about "spaces" – allow for drawings and visualisations which can stimulate ideas, but they are not meant to be very precise.

Nevertheless many materialisations in mathematics are more than visualisations with heuristic value. The process of routine calculating, solving an equation, problem-solving up to and including proving, essentially use materialisations.

The speciality of mathematical representations, through which they become more than mere drawings or written speech, is the following: There exists an developed set of rules for transformation, according to which certain representations can be transformed into others, sometimes in a certain sense equivalent ones. Doing mathematics is for long stretches an interplay between representing and transforming – from calculation in primary school, solving equations, up to the construction of proofs in university; it is an interaction between a thinking human and a mechanically transformed graphical representation. The transformation can be done by the human him/herself or by a real machine, where the human only has to give the command. Charles S. Peirce described mathematics as "diagrammatical thinking". More precise: "It is not by a simple mental stare, or strain of mental vision. It is by manipulating on paper, or in the fancy, formulae or other diagrams – experimenting on them, experiencing the thing" (Peirce, 1966: IV.86; see also Fischer, 1984).

I summarise the above argumentation as a "definition of mathematics": Mathematics is the material, symbolic representation of abstract issues, not immediately conceivable by natural senses, with the potential to transform the representations according to rules. "Symbolic" means that the representations need interpretations, that they are not self-explaining (as is every representation in principle). I know that this definition does not comprise all aspects of mathematics, but I find it useful for explaining the social impact of mathematics.

Some words about the computer. In addition to representations on paper (and screen) the computer can perform the transformations which otherwise are done by interaction between the human and (e. g.) the graphical representation. Despite this, interaction between human and material representation remains necessary also in the case of the computer, in which the "user interface" is graphical. The computer is a representation of abstracts, but simultaneously it requires new forms of representation, namely for the "user interfaces" (programming languages etc.). My guess is that in the future the generation and invention of new forms of representations (notations) will take place more frequently in mathematics than in the past, when this did not always occur in every century and the task of mathematics was to some extent to explore the potential of an existing notation. It may be that in the future the invention of notations will be the genuine creative contribution of mathematics.

I do not want to say much about the importance of forms of representation for mathematics itself. It is obvious that representations of numbers are of crucial significance for calculation-algorithms. For me it is equally obvious that the historical evolution of calculus was highly influenced by specific notations and the algorithms connected with them: representations of numbers, the four basic calculations, notation with variables, tables for special functions. This explains for instance the importance of approximation by functions which can easily be manipulated with respect to differentiation and integration. The computer has changed the situation radically; we now live in a period of transformation.

Note: My view of the relationship between mathematics and matter has nothing to do with the concept of mathematics offered by dialectical materialism, according to which mathematics arises from the material world by abstraction. My concept is in some sense the reverse: Mathematics is applied matter, whereby the relationship is symbolic, that is, mediated by humans.

THE SOCIAL RELEVANCE OF MATERIALISATION

The main focus of the present article is not the relevance of modes of representation to mathematics itself, but their relevance to society. More precisely: *How do mathematical materialisations provide a basis for a rule-oriented society?*

Firstly, there is an immediate relevance, which is important even for the individual: Materialisation gives reality to the abstract; it supports concentration on the abstract and thereby the process of abstraction – namely the disregarding of certain aspects. Nobody has ever seen the "sevenness" of a set; by writing down the digit "7" the sevenness becomes more real. At the same time the notation makes it easier to forget whether 7 apples or 7 pears are meant, or something else. Another example: The relation between the place of a stone which has been thrown up and the time span from the start is an abstract, which can be made more real by drawing the graph of the corresponding function or by writing down its equation. At the same time the person who has thrown the stone "disappears" from consideration; it becomes uncertain whether a stone was thrown or some other object or whether it is another action altogether. Similar points can be made about the visualisation of a calculation procedure by means of a formula and of a system of linear equations by means of a matrix, about the distribution of marks on an examination by a boxplot diagram or about a complex work process with a flow diagram etc. (see Fischer, 2003).

The forgetting of aspects, which are disregarded in the course of abstraction, is important for decision-making. A person who wants to take everything into account never finishes and therefore cannot make (any) decisions. The mark that grades the achievement of a student, the threshold value which fixes the noxiousness of some toxic substance (as a number), are all materialisations of abstracts which disregard many aspects, but they are the basis on which decisions are made. Whether reality is constructed by such materialisations, that is, whether "achievement" or "noxiousness" is defined, or whether reality is described, is a philosophical question which can be discussed in each special case. Whatever the position may be, something becomes more real by materialisation.

The effects of materialisation described above are relevant even for the individual, as already mentioned. But they gain special importance if social systems, collectives are concerned, as subjects as well as objects. This is especially valid if these collectives are large (in number): states, large organisations, whole societies. Firstly, abstracts are relevant for large collectives – complexity generates phenomena which cannot be recognised directly by the senses and are therefore abstract. Secondly, materialisation supports the formulation and implementation of rules and mechanisms which hold together these social systems. And thirdly, putting the abstracts in concrete terms relieves the collective concentration of many people, which is necessary for processes of negotiation among them. (Think about the kinds of questions which can be presented for voting.)

Mathematics fosters concretisation by its materialisations. Beyond its relevance to the individual, matter is that entity about which we have the highest common certainty, and to relate abstracts with matter increases (at least the feeling of) collective certainty.

Some examples of abstracts which are relevant for large systems:

– How well are we doing in our country? Very abstract! Restriction: How well are we doing economically? One (mathematical) answer is *gross national product*, materialised in numbers, tables and graphics. One knows that this answer is inappropriate in many respects, many aspects are disregarded – on the other hand we would not be in the position to negotiate about certain questions, e.g. "Which public social supports are possible and fair?" without this kind of construction.

– What should be the contribution of individuals to common expenses? An answer is given by a *tax system*, materialised in tables and formulas.

– How shall the costs for work by machines, driven e.g. by electricity, be distributed? Here physics helps with the concept of *"energy"* – an abstract which is concretised by numbers and formulas.

– How do we concretise the claims for commodities and services? The materialisation: *money.*

All these abstracts and their materialisations are the basis for social rules and for communication about these rules. It would not even be possible for certain phenomena to become a common public issue without materialisation. The abstracts alone – even if they were designated linguistically – would be too fleeting. Materialisations therefore support *mass communication.*

Mathematics thereby contributes to the (communicative) stabilisation of social systems, including the society of the whole world. Matter serves as a grip from outside. The security of this grip is bound to the validity of the connection between the abstract and its materialisation, an issue which has to be discussed in each specific case. The grips from outside provided by religion or dynasties have been replaced or at least supplemented by matter-based grips. It is true that monarchs used mathematics as a basis for their rule – in fact important areas of mathematics arose from this kind of use – but in feudal regimes mathematics never was so important as it is in (econonomo-)democratic societies.

CONSCIOUSNESS OF SOCIETY

I stated above that the rule-oriented society is unconscious and I also said what consciousness of society is *not*: The sum of individual "consciousnesses" or common opinion and knowledge. What then should it be? I develop the concept by starting with the philosophical concept of individual consciousness as (a process of) self-observation. Consciousness of the individual in this sense means that at the first level the instinctive performance of life, namely that which runs unconsciously, is observed by the individual him/herself. The observation is not restricted to this level; it comprises the process of observation itself. In any case, in order to perform this self-observation, the individual has to put her/himself outside her/himself, look at her/himself, especially at the border of the self. In order to do this, self-difference and self-distance are necessary.

By the mere act of observing, consciousness calls the unconscious operation into question. For the statement (as a result of observation) that something is this or that implies the possibility that it could be otherwise. Metaphorically spoken, consciousness causes pain by this calling-into-question. It is forever the continual expulsion out of the paradise of instinctive living.

How can this concept of consciousness be transferred to social systems? By analogy, the consciousness of a social system means that the system itself, its operation, its boundaries, are objects of observation by the system. The observation is done by the members of the system. The observation may be done by all members, but if the system is large, especially if face-to-face-communication is not possible for all, it is effective to specialise: Only some members of the system will concentrate on observation. To do this and in order to establish a distance to the system they should to some degree be relieved from the usual running of the system. This is done in training groups; organisations sometimes employ consultants to construct an enlarged system with more distance within itself. With respect to whole societies the job of observation is assigned to the (social) sciences, mass media and the arts, which are granted special liberties.

The existence of such instances for observation does not guarantee a consciousness of the social system, at least if there is a democratic claim. Additionally discussion by the whole system about the reports of the observers is necessary. That means that there must exist organisational precautionary measures to support the members of the system so that they can take note of and react to the offered observations. For such common listening and discussion the specialisation, i.e. the bringing some observers into prominence, is helpful as well, at least in large social systems: It fosters collective concentration. In such a discussion conflicts are inevitable, because blind spots and taboos can be made explicit. Even collective consciousness can cause pain.

Consciousness of social systems may neither be identified with a special subgroup – that of the observers – nor with certain contents – the observation reports. Rather it is the (permanent) process, which runs with the participation of observers and all members of the system.

One problem when conceptualising consciousness of social systems in the way I do is that of the existence of the object of observation, namely the social system itself. In fact the social system is a construction, and the process of constructing takes place just by "observation" and discussion about the results, by which the concepts of the respective social system are modified and developed further. Observation reports are to some extent proposals about how the system should be or should not be; the discussion is the site of decisions of the system about itself – for example about the question who belongs to the system and who does not. The discussion necessarily leads to a de-construction of the construction presented by the observers, because – at least some – members will protest against a definition of the whole system which to some extent restricts their freedom. In each specific case it is open whether the observations of the observers become "official" or not; in the long run they are de-constructed because no human social system can be fixed forever.

(The problem of the (non-)existence of the object of observation as a stable entity exists in principle also for individual consciousness. But because of the existence of an individual material body it is not so obvious.)

I now come to the following definition: Consciousness of a social system is an interplay of constructing holistic concepts of the system itself by observation, and a subsequent deconstructing of these very concepts. At least in large social systems the task of constructing is assigned to subgroups, whereas all members are qualified for deconstruction (see also Fischer, 1998, and Fischer, 1994).

What now are, in this connection, the opportunities offered by mathematics? They seem to be obvious. By objectivation – supported by materialisation – of the abstracts mathematics can contribute to constructing the wholeness (of the social system). Simultaneously collective concentration on these very abstracts is made easier, so that it is possible to negotiate about the abstract constructions and the wholeness can be deconstructed.

As a rule the constructive-stabilizing impact of mathematics dominates. In order to provide a contribution to the process of deconstructing, one has to recognise the limits of special mathematical concepts (and their materialisations), which requires a specific critical competence with respect to mathematics. Emotionally a basic distrust of mathematics is necessary – besides the basic trust which is also necessary to maintain the life of the social system. In order to sharpen the critical view and to generate distrust I will in the final part of the paper deal with an incompatibility of the above-defined conception of social consciousness with some principles of mathematics.

DIALECTICAL AUTONOMY

Consciousness of a social system requires that elements of the system, possibly individuals, claim to view (and thereby construct) the whole. For this they need *autonomy*, whereby autonomy should have two components. The first one is usually identified with autonomy generally: independence from the "rest of the world", at least from the system one is part of. It is the competency to set the rules one has to obey oneself, according to the Greek root of the word "autonomy". This component is necessary to establish distance with the observed system (to which one belongs). The second component is in a certain sense the opposite of the first: To be interested in the whole system, to have the ability to recognise phenomena of the system of which one is a part, even if they are not obvious. This requires one to be near the system. And it has also an ethical consequence: to be ready to take over *responsibility* for the whole system. The first component stresses the possibility to stand alone with one's own rules; the second stresses the requirement to be bound to the whole system.

The adult human is confronted with corresponding requirements. One expression of this is I. Kant's categorical imperative: to set the rules for oneself in a way that they are appropriate for the whole system. One is allowed to set one's own rules, but one has to do it with regard to the whole of which one is part.

(It would be interesting to derive criteria for the capability for autonomy of regions, with the claim that autonomous regions (states) have to make a contribution to the process of consciousness of the whole world. At least they should then have the e.g. structural and economical capacity to communicate with the rest of the world.)

The autonomy of part of a system as described above – I call it "dialectical autonomy" – requires the potential to view the whole of which one is a part, to observe it, to make designs, to suggest alternatives etc. I therefore add to the well-known systemic principles, namely "The whole is more than its parts" and its stronger version "The whole is more than the sum of its parts", the "Main Theorem of Dialectical System Theory": "The autonomous part is not less than the whole" (see Fischer, 1994). Obviously there is a logical contradiction. It is the expression of a dialectical relationship between part or element on the one hand and the whole on the other hand. It is a relationship according in which neither side can be reduced to or is subject to the other, a relationship as it is expressed in the well-known sentence of K. Marx, that views the human as the "ensemble of social relations" (Marx, 1958: 6, see also Kuczynski, 1987). A main point in the present context is that a dialectical relationship between parts and the whole is a necessary prerequisite for consciousness of a social system.

THE IRREFLEXIVITY OF MATHEMATICS

The crucial fact now is that it is an important principle of *mathematics to have a non-dialectical relationship* of parts and the whole. This can be seen, for instance, in G. Cantor's "definition" of a set as the "summing up of well-distinguishable objects [...] to a whole" (Cantor, 1895). This definition implies: The elements exist before they are summed up and independently from the product of summing up, the whole set. The collecting, the building of the set, is a constructive act from outside. The whole set is not included in any one element (not even as an idea). Symbolically:

$$M \subseteq x \in M$$

is not allowed (see Meschkowski, 1966: 24). It is even not allowed that the set be contained in a proper subset.

I want to illustrate the situation by still another mathematical concept, namely the concept of *function*. This concept requires two separate entities: a rule for assigning and the area of objects to which this rule is applied, usually called the domain of the function. These entities must be separate; *it is especially not allowed that the elements of the domain define the rule.* Certainly, by introducing additional parameters and thereby enlarging the domain it can seem as if this were possible, for instance:

$$f(x) = x^2 \qquad \text{is enlarged to} \qquad f(x,n) = x^n$$

But such an enlargement will never be exhaustive; an additional (meta-)rule will always be required which is not defined by the elements of the domain. These are in some sense "subject" to the rule. M. Otte describes this situation as that of a "separation of relations from the related" (Otte, 1993: 402, translation by R. F.). In a hierarchy of logical types functions are on another level than the elements (to which a function is applied) and may not be mixed up.

One now can establish an analogy to the usual concept of organisation, especially in business administration. An organisation is fixed by its structure, its rules etc., but it is not allowed that this structure should be under total control by the members of the organisation. At least at the beginning a designer, an author of the constitution, an authority from outside is necessary. Even democratic organisations need a basis, maybe general principles, which are not subject to the will within the system. Any organisation, according to the usual conception, needs an invariant kernel of structure and rules.

Especially the question of how to govern organisations is usually answered on the basis of this idea, even if one believes to have left rigid bureaucratic concepts of organisation behind. The "logic of function" seems to be compelling: How else should one be able to govern a (social) system unless some components of the system are fixed? And moreover: How else could the identity of a (social) system be constituted, if not by abstraction from the elements towards an invariant mechanism of processing the system? And should not mathematics receive the merit that by its way of thinking this abstract invariant can be named and perhaps even represented by appropriate concepts?

But there exist alternatives to the dominant concept of organisation. In the field of designing and managing social organisation ("organisational development", "theory of management") there have for several decades existed efforts to invent and try out new understandings of organisation and new models of governing which are not based upon the "logic of function" or upon the separation of structure and elements. One speaks about "learning systems", about "evolutionary management", about "coupling up" and "irritating" instead of "governing" etc. (see Beer, 1986; Ulrich & Probst, 1984). The aim is, with greater or lesser radicality, to give "more rights" to the system, to view it not only as subject to the will of a "governor". Nevertheless the way to accomplish this is sometimes simply to cancel the elements, to concentrate on the structure and integrate the arising of will into the structure. The dialectical approach however retains the elements and ends up at the question of whether "equal rights" of elements and structure can be realised practically.

As already pointed out above, for mathematics a dialectical relationship of elements and structure is not comprehensible; mathematics is "irreflexive". If mathematical thinking is related to the wholeness of social systems and if it is not only restricted to solving well-restricted problems, mathematics can hinder self-reflexivity and thereby consciousness. This is a real danger, not because mathematical models of social or economical development are so important, but because the principles of mathematics are deeply anchored in our individual and social world of thinking; even people who refuse mathematics on a superficial level are infected. There is a strong impact of mathematics onto our social fantasy.

On the basis of a concept of humans, according to which individuality and sociality are related dialectically, mathematics is inhuman. But one has to add that almost no (scientific) discipline makes contributions to dialectical organisation. So mathematics is not in bad company. Moreover: In its fundamental considerations it gets to the point of the basic assumptions of the dominant disciplines. Going beyond this defensive diagnosis, I dare to claim still more: If mathematics is used properly, it can foster dialectical organisation and thereby collective consciousness.

KEEPING VS. OVERCOMING

The description of a situation by means of mathematics can be used for two different goals:
– to keep the situation, to stabilise it and to legitimise existing circumstances, or
– to call the existing situation under question, to change, to overcome it.
The first goal is usually aimed at, when the organigram of a firm is drawn. The second goal is completely different. I want to illustrate the difference by using the model of the *sociogram*. Sociograms are tools to describe the situation of a (social) group. One asks, for example, each member of the group with which other members (s)he wants to cooperate. A graph with vertices symbolizing the members of the group and directed edges from each member to the desired partners of cooperation would be a sociogram. This tool was "invented" at the beginning of the 20th century by the Austrian psychologist L. Moreno (1974). It was developed further with various concepts and even theories (e.g. "star" of a group, or "subgroup" etc., see for example Seidman, 1985). Additionally there have been attempts to find out general theorems about structures and evolution of groups by means of a series of experiments with groups.

But the motive of Moreno was different: Using the sociogram he wanted to hold a mirror up to the members of the group so that they could recognise their situation and change it – he spoke about "micro-revolutions" – leading to a situation where the sociogram is no longer valid. Sociograms should therefore not only be a means of description, but also a means of intervention; a means which fosters the process of consciousness of the group, including de-construction. To formulate it a little bit exaggeratedly: A sociogram has performed well if it is no longer valid in the future.

This situation is completely different from the usual situation when mathematical models are constructed in natural, technical or economical sciences. In all these cases the modeler hopes that the model will also be valid in the future, such that prognoses can be made; at least some meta-structure should be permanent. But if one uses a sociogram in the sense of Moreno, no prognosis can be made – not even that the situation will change.

Now I am at the point where I can say: Mathematics is able to contribute to the process of consciousness of a social system, namely if it is used as an element of operation in an interplay of keeping and overcoming; an interplay which corresponds to construction and de-construction. I go a step further: By its property as a powerful means of representation, which brings to light structures very precisely and clearly — in some cases by transforming the representations by letting consequences come to light — mathematics can contribute to changing these very structures; namely by provoking decisions which lead to change. Its "decidedness" and precision serve to make everything acute, so that the limits of given conditions might become obvious. Simply with this potential mathematics can, if it is not used dogmatically in order to legitimate the given, contribute to its self-overcoming. In my opinion, as in no other discipline, mathematics has the potential to overcome itself.

Do we, by these deliberations, go beyond the borders of what mathematics is? If mathematics is simply the sum of its models, theories and concepts, then the answer is "yes". If mathematics comprises its development, locally in the work of a single mathematician as well as globally in its historical evolution, then the answer is "no". Because the history of mathematics itself constantly shows situations of overcoming.

The question is what we want that mathematics should be. Especially mathematics education, from elementary school up to university, has to decide. We can work to form the way into an unconscious rule-oriented society. Or we can contribute to more consciousness of the society.

NOTES

[46] First published in German: Technologie, Mathematik und Bewußtsein der Gesellschaft. In G. Kadunz, G. Ossimitz, W. Peschek, E. Schneider & B. Winkelmann (Eds., 1998) Mathematische Bildung und neue Technologien. Stuttgart: Teubner, 85-101.
I am indebted to James Edinberg for improving the text with respect to English language.

REFERENCES

Bammé, A., Baumgartner, P., Berger, W., & Kotzmann, E. (Eds.) (1987). *Technologische Zivilisation. Eine Einführung*. München: Profil.

Bammé, A. (1993). Der EAN-Code. Auf dem Weg zu einem Warenwirtschaftsverständnis jenseits von Kapitalismus und Sozialismus. In W. Berger & A. Pellert (Eds.), *Der verlorene Glanz der Ökonomie*. Wien: Falter, 309-330.

Bammé, A., & Fleissner, P. (1994). Was hält die Welt zusammen? Gesellschaftliche Synthese durch Technologie oder Ökonomie? *Kurswechsel*, 1/1994, 63-82.

Beer, S. (1986). Recursions of power. In R. Trappl (Ed.), *Power, autonomy, utopia. New approaches towards complex systems*. New York: Plenum.

Cantor, G. (1895). Beiträge zur Begründung der transfinitiven Mengenlehre. *Mathematische Annalen, 46*, 481-512.

Fischer, R. (1984). Offene Mathematik und Visualisierung. In R. Fischer (Ed., 2006), *Materialisierung und Organisation. Zur kulturellen Bedeutung von Mathematik*. München: Profil, 223-256.

Fischer, R. (1994). Drei Paradigmen systemischen Denkens. *Wissenschaftliche Blätter/Angewandte Ökologie der Wissenschaftlichen Landesakademie für Niederösterreich*, 1/1994, 38-40.

Fischer, R. (1998). Wissenschaft und Bewußtsein der Gesellschaft. In L. Gubitzer & A. Pellert (Eds.), *Salbei und Opernduft. Reflexionen über Wissenschaft*. Zeitschrift für Hochschuldidaktik, 3/1998, 106-120.

Fischer, R. (1998a). Technologie, Mathematik und Bewußtsein der Gesellschaft. In G. Kadunz, G. Ossimitz, W. Peschek, E. Schneider & B. Winkelmann (Eds.), *Mathematische Bildung und neue Technologien*. Stuttgart: Teubner, 85-101.

Fischer, R. (2003). Reflektierte Mathematik für die Allgemeinheit. In L. Hefendehl-Hebeker & S. Hußmann (Eds.), *Mathematikdidaktik zwischen Fachorientierung und Empirie. Festschrift für Norbert Knoche*. Hildesheim: Franzbecker, 42-52.

Kuczynski, T. (1987). Einige Überlegungen zur Entwicklung der Beziehungen zwischen Mathematik und Wirtschaft. In W. Dörfler, R. Fischer & W. Peschek (Eds.), *Wirtschaftsmathematik in Beruf und Ausbildung. Vorträge beim 5. Kärntner Symposium für Didaktik der Mathematik*. Wien: Hölder-Pichler-Tempsky, 145-166.

Luhmann, N. (1986). *Ökologische Kommunikation*. Opladen: Westdeutscher Verlag.

Marx, K. (1958). Thesen über Feuerbach. In K. Marx & F. Engels, Werke, vol. 3. Berlin: Dietz.

Meschkowski, H. (1966). *Einführung in die moderne Mathematik*. Mannheim: Bibliographisches Institut.

Moreno, J. L. (1974). *Die Grundlagen der Soziometrie. Wege zur Neuordnung der Gesellschaft*. Opladen: Westdeutscher Verlag.

Otte, M. (1993). *Das Formale, das Soziale und das Subjektive. Eine Einführung in die Philosophie und Didaktik der Mathematik*. Frankfurt: Suhrkamp.

Peirce, C. S. (1958-66). *Collected papers* (ed. by C. Hartshorne, P. Weiss & A.W. Burks). Cambridge: Harvard University Press.

Seidman, S. B. (1985). Models for social networks: Mathematics in anthropology and sociology. *The UMAP-Journal, 4*(2), 19-36.

Ulrich, I., & Probst, G. J. B. (1984). *Self-organization and management of social systems*. Berlin: Springer.

AFFILIATIONS

Roland Fischer
Fakultät für interdisziplinäre Forschung und Fortbildung,
Alpen-Adria-Universität Klagenfurt

OLE SKOVSMOSE

STUDENTS' FOREGROUNDS AND THE POLITICS OF LEARNING OBSTACLES [47]

In 1954, Hendrik Verwoerd made the following statement in his address to the South African Senate:

> When I have control over Native education I will reform it so that the Natives will be taught from childhood to realise that the equality with Europeans is not for them [...] People who believe in equality are not desirable teachers for Natives [...] What is the use of teaching the Bantu mathematics when he cannot use it in practice? This idea is quite absurd. (Verwoerd, quoted from Khuzwayo, 1997: 9)

According to Verwoerd and the apartheid regime, the paramount task was to make sure that blacks were prevented from climbing the social ladder. Being excluded from mathematics also meant being excluded from the possibility of advancement in society.

In other words, learning obstacles for a certain group of students can be established in explicit ways such as being subject to an absurd policy. Here we are a long way away from the epistemic notion of learning obstacles, analysed in terms of students' preconceptions, if not misconceptions, of some mathematical notions and ideas. The epistemic interpretation of 'learning obstacle' is *not* the only one possible. However, processes of exclusion in education can be dressed up in such a way that their political dimension becomes hidden and ignored. It could appear that exclusion is not imposed on students. Instead, exclusion may appear as a consequence of some students' so-called low achievement.

When mathematics education operates as part of social mechanisms, providing or justifying certain forms of inclusion or exclusion, it comes to serve as a gatekeeper. According to Volmink (1994: 51-52),

> to deny some access to participate in mathematics is [...] to determine, *a priori*, who will move ahead and who will stay behind.

This statement can be read as a dramatic description of the role of mathematics education as institutionalising a distinction between those who are included and those who are excluded. Bourdieu (1996) refers to a 'state magic', by means of which the state assigns some authority to a group of people, by referring to that group's good performance in school, and in particular to the performance in mathematics.

U. Gellert, E. Jablonka (eds.), Mathematisation – Demathematisation: Social, Philosophical and Educational Ramifications, 81–94. © *2007 Sense Publishers. All rights reserved.*

To me the notions of inclusion and exclusion are as relevant for the discussion of mathematics education as the notion of achievement. Is it initially high or low achievement that produces social inclusion or exclusion? Or do socio-political processes of inclusion and exclusion manifest themselves in schools and in the mathematics classroom as high or low achievement? Needless to say, the relationship between high/low achievement and inclusion/exclusion is more complicated than that, but I find it essential to consider carefully how social processes of inclusion and exclusion might appear in the mathematics classroom. In this way, it might be possible to discuss the politics of learning obstacles.

I am going to consider such issues from a conceptual perspective. In particular, I am going to draw attention to the notion of *each student's foreground* as a conceptual construct that might facilitate a discussion of the 'politics of learning obstacles'.

The introduction of the notion was inspired by some observations I did years ago of seven-year-old students doing mathematics. At the beginning of each lesson, the teacher carefully explained the new task. Following this, the students had to do some exercises from the textbook. Suddenly several students seemed completely occupied in their own work, and when I tried to engage the boy sitting next to me in a conversation he simply ignored me. I soon realised the reason. After each student had completed the exercises, they went to the teacher's desk, and the teacher commented in an encouraging way about what each student had done.

However, there was another agenda in operation. As the students lined up, a public stratification took place: Who was first, second, third? A competition took place amongst the students, at least among some of them.

I also observed a group of girls who seemed to be ignoring this rush. They did address the exercises, but they worked in their own pace and with a lot of small interruptions. They seemed, somehow, to have renounced speed, realizing, perhaps, that whatever they did they would never become among the first in the line. It was common knowledge that sometimes Peter (who I was sitting next to) was the first, sometimes Anna, sometimes Maria, and a few times a fourth one, but the candidates for winning the game were a minority.

What could one do better than establish one's own priorities? The group of girls had started the first problem, but soon they needed to erase a wrong number, and spent time choosing the proper eraser. Should it be the one with a smell of strawberry? Pencils had to be sharpened; comments and smiles had to be exchanged; and in these ways time would pass. Somehow, they protected their integrity very well. Had they, instead, struggled in vain to get first in line, lesson after lesson, their self-esteem might have been affected.

To understand their approach to mathematics I explored how they interpreted their opportunities. I found that they constructed their 'foregrounds' as expressions of what they experienced as their opportunities, and that these foregrounds set the conditions for their engagement in, as well as their resistance towards, mathematics.

Originally I developed the construct of 'foreground' from a philosophic perspective and used it in the interpretation of learning processes taking place in

Danish classrooms (Skovsmose, 1994; Alrø & Skovsmose, 2002). However, it has turned out to be useful in other, different, educational contexts:
- for a co-operative project conceptualising and establishing a mathematics education with an explicit concern for democracy (running since 1994) between universities in Denmark and South Africa – discussing educational consequences of the apartheid regime's policy and of the damage to opportunities in life that this regime caused for black students
- for exploring further, also since 1994, the ethnomathematical research programme in an extensive co-operation with researchers from Brazil.

Thus, I do not consider the construct 'foreground' to be relevant only when we consider the situations with reference to which the construct emerged. Instead, I consider the construct relevant for any discussion of learning that includes a discussion of learning obstacles.

PRIVATISING OR POLITICISING LEARNING OBSTACLES?

Let me illustrate what I mean by *politics of learning obstacles* by summarising one aspect of white research on black education carried out during the apartheid past of South Africa. Interpretations of achievement in school was a big issue, as particular interpretations could help to 'explain away' the brutality of the apartheid regime. Racism was an all-embracing category. Both 'classic' and 'progressive' racism argue that it is not the school that has to be blamed for the weak performance of the black child, but that the child brings the cause for his or her weak performance into school.

In the basic assumption of *classic racism*, performance in school was accounted for by referring to certain 'facts'. That black children did not perform as well as white children had to be understood in terms of biological structures, established thousands and thousands of years back in time. Certainly, such an explanation provided a solid distance between the apartheid regime and the causes for what was observed in classrooms.

In particular, black children's learning obstacles had nothing to do with the school structure, and certainly nothing to do with apartheid politics. These obstacles had to be found in the black children themselves. They bring their own deficiency along with them, right into the school. The children are inevitably linked to their own poor performance, which is just a different expression of their colour of skin. Thus, the political dimension of school performance is efficiently explained away.

In *progressive racism*, the idea that social aspects play a fundamental part in a person's intellectual and emotional development, and not the biological framing, did lead to new priorities within white research on black education. Instead of searching for a biological explanation of the weak performance of black children, social factors, intrinsic to 'black culture', could be identified. In his study of white research carried out during the apartheid era of the Orange Free State University, Khuzwayo (2000) has uncovered what such research included: How to explain observations of black children's weak performance in school? One suggestion was that reasons had to be sought for in their social background.

In one study from the Orange Free State University, an explanation was suggested in terms of family traditions and, in particular, in terms of the dominant role of the father in the black family. According to this study, this aspect of the black family helps to explain that creativity, and also the mathematical creativity of the black child, was eliminated. Thus, family structures became a main factor in the explanation of the black child's performance in school. The problem is thus to be found in the cultural background of the child. In other words, black culture (and therefore not any suppressing white culture) produces the learning obstacles for black children.

Black children's problems in school are established beforehand and are not to be located in the school structure. The black children themselves bring their learning obstacles to school. The best the school can do is to compensate for such cultural deficiencies. According to Bourdieu's (1996: 10-11) formulation, racism establishes categories of perception and forms of expression,

> which suppress or repress the social dimension of both recorded and expected performances and [...] dismiss any questioning of the causes.

Many deficiency theories (theories of the deprived child) follow the approach of racism in explaining away the socio-political dimension of school performances, by *privatising* and *personalising* the causes of such performance.

Bopape (2002) has made a study of mathematics education in the most desolate parts of South Africa, and he has showed me what a school might look like; broken windows; doors and all electric installations missing. There was a hole in the roof – maybe the tiles had been removed by somebody who found that his or her house needed the tiles a bit more than the school building did. When it was raining, the children had to get away from this part of the classroom. The classroom would be either too hot, or too cold, or too wet. It looked like a place where teacher and students would meet having a shared intention to leave this ugly place as soon as possible. What seems to be the most obvious learning obstacle to the children in this school: their colour of skin, their dominant father, or the hole in the roof?

What does this mean for mathematics education? It is all too obvious: When we enter the classroom, the first thing that hits us is the evident physical learning obstacles. The learning obstacles are right there in front of our eyes, and on top of our heads. To me, this hole in the ceiling, not referred to in much research in mathematics education, calls any deficiency theory of the child into question (see also Ginsburg, 1997; and Gorgorió & Planas, 2000, 2001).

How could it be that this hole in the roof has not been seriously addressed by mainstream research in mathematics education? Learning obstacles can be looked for in the actual situation of the children and with respect to the opportunities which society makes available for the children. The actual distribution of wealth and poverty includes a distribution of learning possibilities and learning obstacles. This distribution is a basic political act. Paying attention to this means re-establishing the politics of learning obstacles. Civil and Planas (2004: 12) address the notion of learning obstacle in a similar way:

> [L]earning obstacles are not related to students' preconceptions and misconceptions of some mathematical ideas, but to forms of excluding some groups from mathematics education.

And they emphasise that learning obstacles go "much deeper than classroom dynamics" (p. 12).

During the rest of this article, I shall try to explore in greater detail the nature of learning obstacles by considering the notion of *each student's foreground*. By means of this notion I shall try to illustrate that learning obstacles can take the form of a ruined foreground, and that ruining the foreground of a certain group of children is a socio-political act. That a foreground is spoiled does not mean that there is no foreground, but that the foreground appears to be without attractive and realistic opportunities. Learning can, to a student with a shattered foreground, appear pointless: Why should I try to learn this? Why should I pay attention to mathematics? I cannot see any meaning in this. A ruined foreground does not easily support hopes. However, before getting to these ideas, I want to take a look at the notion of background.

MEANING AND BACKGROUND

Bishop (1990) refers to a mathematics textbook that contains the following problem:

> The escalator at the Holborn tube station is 156 yards long and makes the ascent in 65 seconds. Find the speed in miles per hour.

Working with this problem has a very different meaning to children in London than to children in Tanzania who, in fact, have been introduced to this textbook. In the first case the exercise can be seen as an attempt to provide mathematical exercises with some meaning (although I doubt how successful the textbook author has been in this case). However, when the textbook is used with children in Tanzania, as part of a programme imposed on them by the British colonial officers of the time, the same exercise can be seen as cultural imperialism, and mathematics education can be characterised, as Bishop has done, as a weapon of cultural imperialism.

Meaning has been discussed with reference to the notion of culture. Meaningful mathematics education has been sought for by relating classroom activities and possible contextualisations of exercises to the students' culture and background. The escalator at the Holborn tube station may bring some meaning to some English children. Bringing the students' *cultural background* into the classroom as a resource for contextualisation seems relevant for bringing meaning to the mathematics classroom. To draw on students' background appears to be a simple and obvious idea. Nevertheless, I am not sure that we are dealing with a 'simple truth'.

Culture can refer to many things, and cultural background can operate in different ways in the classroom. Culture can refer to tradition and folklore. But when expressed by the apartheid system, a notion like 'Zulu culture' comes to operate in an oppressive way. It could constitute a trap. An 'appreciation' of Zulu culture might be associated with the assumption that people belonging to this culture stand outside the Western development. The Zulu culture could be picturesque and include, for instance, dances with traditional weapons. The notion of culture could get a negative connotation, referring to people 'out there' and 'down there', supporting the view that people with such a 'different' culture had better stay in their homelands.

It appears problematic if an appreciation of culture results in a constant appreciation of traditions. Culture is changing and developing, it includes new elements, good and problematic ones, in a complex mix. 'Culture' is a contested concept. As a consequence, it is not a straightforward act to search for meaningful mathematics education in the cultural background of the students.

In much ethnomathematical literature the intimate connections between culture and mathematics has been emphasised. D'Ambrosio (2001) explains that by *ethno* he refers to the fact that mathematics is acted out in many different ways in different cultures and by different groups. Mathematics is always socially embedded. The ethnomathematics perspective has had implications for the practice of mathematics education: It has been emphasised that it is important to consider the background of the students when we try to constitute meaningful mathematics education.

I certainly agree with the point that making mathematics education meaningful is essential. I also agree that meaningfulness has to do with the cultural background of the students, but I would argue that meaningfulness has much to do with another dimension as well, namely the foreground of each student. To me 'cultural background' should not remain the only key notion when meaningfulness in mathematics education is discussed.

LEARNING AS ACTION

I find it problematic to try to explain the performance of somebody by, first of all, referring to the background of the person. This could be any kind of performance, and also (high/low) achievements of students in a mathematics classroom. 'Referring only to the background' is a strategy by means of which the political nature of learning obstacles can be eliminated. If, instead, we try to explain the performances in terms of the background, the here-and-now situation, and the foreground of each student, then the political nature of the learning obstacles becomes more obvious.

By the *foreground* of a person I understand the opportunities, which the social, political and cultural situation provides for this person. However, not the opportunities as they might exist in any socially well-defined or 'objective' form, but the opportunities as perceived by a person. Nor does the background of a person exist in any 'objective' way. Although the background refers to what a person has done and experienced (such as the situations the person has been involved in, the cultural context, the socio-political context and the family traditions), then background is still interpreted by the person. Taken together, I refer to the foreground and the background of a person as the person's *dispositions*[48] (and for simplicity I include the meeting point between foreground and background, the present situation, in the foreground).

These dispositions need in no way be homogeneous entities. Dispositions can incorporate conflicts and contradictions. A person can conceptualise different sets of aspirations, *i.e.*, different foregrounds, and his or her background can be structured by conflicts. And, certainly, a person's aspirations need not square with his or her background.

Dispositions should not be taken to be characteristic of only an individual. It makes sense to talk about the dispositions of a group with shared background and foreground. Dispositions provide resources for action – not in the sense that dispositions 'cause' actions, but dispositions embody propensities that become manifest in actions, choices, priorities, perspectives, and practices. Dispositions are continuously reworked and remoulded. They are dynamic. They are relational or interpersonal characteristics. They are sociologically structured. This being said, I continue to talk as well about a person's dispositions as those which are reworked and remoulded in the person's interactions with other people, but which the person can draw upon as his or her resource for action.

According to many philosophic interpretations, and mine also, intentionality or intentions-in-action is a defining element of an action. In order to understand a person's action, we have to consider his or her intentions. Behaviourism proposed that actions could be identified with their physical appearance. In this way a certain conceptual framework was suggested for how to explain actions – in terms of events taking place before the actions are carried out.

Several alternative interpretations have been suggested indicating that an action is more than its behavioural appearance. Following the idea of Brentano (1995)[49] it has been suggested that human consciousness can be characterised in terms of its directedness, its intentionality. I propose an interpretation of action that tries to grasp the specificity of action through its intentionality. This makes intention-in-action a crucial construct.

The intentions of a person are not simply grounded in his or her background, but emerge also from the way the person sees his or her possibilities. In other words, I see intentions as rooted in a person's dispositions. Intentions express expectations, aspirations and hopes. Intentions make up a constitutive part of any action, just as actions without intentions degenerate into simple physical movements. Actions become not simply caused by the past, but represent forms of grasping the future. When we want to try to understand how and why a person is acting, then it is important to get an understanding of the person's disposition.

I see learning as action (not all sorts of learning, but some). This is an important idea in the concept of learning that I am considering (see Skovsmose, 1994; and Alrø & Skovsmose, 2002). This notion does not represent a simple description of learning in general, as there are many forms of learning that I cannot describe as action (but instead, in my terminology, a forced activity). Situations where somebody has to learn some routines, like soldiers learning how to go marching in rows, I do not refer to as learning as action. Nor do I consider learning as action, when learning so to say takes place beneath the conscious level of the person, like, without noticing it, adjusting oneself to certain habits when put in a new cultural context. Learning as action cannot be forced upon somebody, but students can be invited into situations where they can be involved in processes of learning as action.

Such learning presupposes that students establish intentions-in-learning, as intention-in-action is a defining element in an action. This means that the students see meaning in what they are doing. Maybe it is even better to think of learning as inter-action: learning means doing things together.

However, a particular situation or a particular way of organising teaching-learning processes can prevent students from acting as learners. This was what I observed in the classroom with students reacting to a public stratification. Some engaged in the activities and put extra effort into what they were doing, whilst others, including the group of silent girls, withdrew their intentions from the 'official' activities of the classroom, and redirected them into 'alternative' activities. In order to interpret the activities of the group of 'silent girls' I had to consider not only their backgrounds, but also what they saw as their opportunities. Their foregrounds might prevent them from putting effort into the designated activities.

I see achievement (low or high) as being related to the opportunities that the school structure and the socio-political context in general make open for the students to perceive as their opportunities. This provides a different interpretation of learning obstacles. Such obstacles have not only to be sought in the historical past of the person, but also with reference to the opportunities that the social and political system makes available for the person. In particular, the apartheid system destroyed the future of black children, and this explains some of their behaviour in school. *When a society has ruined the future of some group of children, then it has also obstructed the incitements of learning.* A ruined future can be the most brutal form of learning obstacle. This interpretation of learning obstacle accuses the society in question. A spoiled foreground is a dramatic learning obstacle, and spoiling a foreground is a socio-political act.

MEANING AND FOREGROUND

In order to establish meaning in education, students should be involved in meaning production, and each student's foreground is an essential resource for this production. I decided to talk about *meaning production*, following a terminology presented by Lins (2001). However, I do not follow the theoretical framework also presented by Lins, as I restrict my use of meaning production to a metaphoric sense, emphasising that meaning is produced and constructed.

Let us imagine that we, as mathematics educators, arrive in a village far away 'beyond the mountains', and we find that in this village there are chickens everywhere. We realise immediately that all kinds of mathematical activities can refer to these chickens, such as counting, selling, buying and cooking. What a perfect situation for making sense of mathematical notions! We construct tasks that relate to counting and selling and buying and cooking chickens. The students will be familiar with it all. This is part of their background. Nevertheless, this need not make sense to students. They may be more interested in, say, pilot's mathematics, although they have only seen an airplane passing by up there high in the sky making a fine white line, signifying that there are many different places to go. When we, as mathematics educators, come to this village we must not only consider the students' background, but we must also consider their hopes and aspirations. We must consider where they want to go. *Meaning not only represents the past. It also represents the present and the future.* Each student's foreground is a principal resource for meaning production.

To me the notion of meaning production emphasises that meaning is not a referential property[50]. Meaning is produced; it becomes an aspect of acts – and not just an aspect of concepts. This is in line with the ideas indicated by Wittgenstein (1953) suggesting an interpretation of meaning in terms of use. Following this direction, I see meaning as an aspect of acts, and meaningful education means that students are invited to engage in meaningful learning acts. Meaning is produced by students, and by co-operation among students and between students and teachers. This production is resourced by the dispositions of the students. Therefore, 'meaning of learning', 'meanings for students', and 'each student's meaning production' must be investigated and interpreted with reference to the dispositions of the students (including their background and foreground).

Meaning production takes place in terms of what the students see as their opportunities, including motives, perspectives, hope and aspirations. It gets its extra fuel from the foreground of each of the students. Meaning production, however, can also be obstructed. If meaning production takes place with reference to the foreground of the each student, then ruining the foreground for certain groups of students creates a real learning obstacle. To a black child in apartheid South Africa, what could be experienced as the point of struggling with mathematics, when jobs which presupposed mathematical skills, like engineering, for instance, were not for blacks? What was the point for many girls in Denmark (or other Western countries) in a not so distant past of concentrating on studying physics and mathematics when jobs, demanding such skills, were apparently for men? Based on her studies, Wedege (1999) claims that the habitus (as interpreted by Bourdieu) of a young woman in Denmark in the 1930-1940s does not automatically encompass a disposition for learning mathematics, nor does it generate a conception of mathematics as a relevant subject. How do children of immigrant parents conceptualise their opportunities in today's Danish (or European) society, and how does that affect their attitudes towards learning? How does an apparent marginalisation of some groups of people affect some groups of students' foreground and therefore their possibilities for being involved in certain learning processes? How does the emerging racism in Western Europe provoke learning obstacles to certain groups of students?

As each student's foreground is one principal resource for meaning production, then ruining a foreground becomes a learning obstacle. This has to be discussed in terms of the socio-political situation of the students. As I see it, by considering the processes of globalisation and ghettoising, we get an idea of the basic condition of learning for many children around the world. Here we find fundamental forms of establishing or ruining foregrounds.

GLOBALISATION INCLUDES EXCLUSION

Globalisation refers to inclusion, but also to processes of exclusion. Castells (1998) discusses the rise of what he calls the Fourth World as an aspect of globalisation:

> This widespread, multiform process of social exclusion leads to the constitution of what I call, taking the liberty of a cosmic metaphor, the black holes in information capitalism. These are the regions of society from which, statistically speaking, there is no escape [...]. (p. 162)

Furthermore:

> The Fourth World comprises large areas of the globe, such as much of Sub-Saharan Africa and impoverished rural areas of Latin America and Asia. But it is also present in literally every country, and every city, in this new geography of social exclusion" (*ibid*, p. 164).

One essential question concerns the possible foreground of each student living within or close by the Fourth World. What aspirations could they have? How do they see their opportunities? What opportunities could be created for them in their process of schooling (if they have the opportunity to go to school)? What can be created within mathematics education? Schools have got a tricky position in the "new geography of social exclusion". This geography determines the foregrounds of the majority of students around the globe. It determines the disposition of groups of people, and as a consequence it structures motives for learning and for meaning production. Many schools are positioned on the borderline between the Fourth World and the informational society. I see much of mathematics education positioned right there in "borderline schools" (Penteado & Skovsmose, 2002; Penteado, 2001).

'Schooling' can be seen not only as a support for entering the network society, but it can also become a gatekeeper, and an 'excluder' from the network society. If students perceive their foreground as ruined, this could easily turn into low achievement, which, in turn, could confirm their exclusion. In this way, schooling can mean a preparation for the dumping of people into the Fourth World. The remarkable statement of Verwoerd thus represents a clear indication of what it means to put people into a ghetto, in this case represented by homelands. The grand idea of apartheid was built upon the notion that people should be separated, and that black people had no role to play in white society beyond the level of unskilled labour. The existence of ghettos of the network society seems to indicate that the informational economy does not need everybody. Only a part of the global population fits into the networking – the rest had better be left in their 'homelands'.

In the days of apartheid, learning obstacles had their obscure significance, but recent processes of globalisation also cause dramatic learning obstacles to many students. What does it mean for each student's aspirations and hopes to be located close to the black holes of the informational society? Castells (1998) refers to the millions of homeless, incarcerated, prostituted, criminalized, brutalized, stigmatized, sick and illiterate persons, who are literally expelled from the informational economy with no functional role to play. From a certain economic perspective, they appear disposable. They are worth nothing as consumers; they have no value as possible human resources for production. Other and sufficient human resources are available. This is one brutal aspect of the informational economy and of the 'new geography of the social order'. Children experience this new order, and many as something highly problematic. This influences their foreground, and therefore their dispositions and motives for learning. To me, a ruined foreground is a learning obstacle. As a consequence, I find it important to discuss learning obstacles with references to globalisation and ghettoising as represented by the Fourth World.

SEARCHING FOR THE FOREGROUND

If we want to understand the learning activities of students and the way they engage in learning processes, it is important to understand what they consider to be their foreground and their possibilities in life.

I do not see any contradictions in assuming, for example, that working with computers and playing with dynamic geometry can be meaningful when we deal with marginalised students living close to the Fourth World. Working with computers may be miles away from their cultural background. But this background does not represent the only essential parameter in their meaning production. Their foreground is important as well. So, an essential question is how activities, say, including playing with dynamic geometry, may capture their hopes and aspirations. It is important that mathematics education provides opportunities, and when these appear to be real opportunities, seen from the students' perspectives, they can become active in their processes of learning.

There is no point in mathematics education celebrating a certain kind of technological culture, or celebrating the network society. There is no point in mathematics education presupposing an appreciation of any 'high-tech' culture. The point is simply that mathematics education should not organise resources for meaning production in such a way that students are parked in a Fourth World.

Valero (2002) suggests a change in interpretations of learners' activities in the classroom. Instead of emphasising the cognitive interpretation, Valero suggests a socio-political interpretation. She refers to a turning point for her reflections taking place during her observations in a school in Bogotá. On one occasion the teachers were missing, and Valero took over the mathematics lesson:

91

> I was supposed to give them the exercise worksheet that they had started to solve the previous session. While many in the class worked, two male students engaged me in a chat. These two boys were supposed to be doing the mathematics. Instead, they looked at the worksheet and laughed. They called me for help, but in reality they were curious to know about my intentions and motivations to be in their school. (Valero, 2002: 490)

The students could not understand why Valero had gone to their poor school and chosen to talk with poor students. The conversation made Valero emphasise that there were reasons to study, and also to study mathematics. Then one of the students answered:

> The only class I would like to pay attention to is English, because I want to get out of this fucking country and to go to the US.

It is difficult to find a clearer expression of the idea that reasons for learning are to be found in the foreground of the students, and that a ruined foreground is a learning obstacle of enormous dimensions. Students may be well aware of this. Learning obstacles are interpreted by students, and the socio-political and economic dimension of the learning obstacles might turn into a student's or a group of students' resistance towards learning. Thus, Planas and Civil (2002: 15) observe:

> If students cannot see any perspective in what they are doing, then we cannot expect any meaningful participation of students.

Alrø & Skovsmose (2002) discuss in detail a group of students who cannot find any motivation with which to engage with the proposed task in the classroom. They are not able to put their intentions in the activities – they find no motives. A learning obstacle becomes acted out as a learning resistance.

It is not possible to claim that mathematics education can fundamentally affect the socio-political processes of inclusion and exclusion. Processes of globalisation are stronger than educational processes. They influence the dispositions of many students, *i.e.*, their motives for learning as well as the learning obstacles they come to face. Nevertheless, I see the task of mathematics education as being to provide opportunities for students. Mathematics education has to realise the foreground of the students as a resource for any meaning production. To look for each student's foreground, and to try to relate learning activities to this, becomes an ongoing challenge to any mathematics education.

ACKNOWLEDGEMENTS

I want to thank Helle Alrø, Bill Barton, Núria Gorgorió, Arne Astrup Juul, Miriam Godoy Penteado, Núria Planas and Paola Valero for critical comments and suggestions for the improvement of this article.

NOTES

[47] This chapter is based on the paper 'Foregrounds and politics of learning obstacles' in *For the Learning of Mathematics, 25*(1), 4-10. A previous Portuguese version of this paper has been published in: J. Ribeiro, M. Domite, S. do Carmo & R. Ferreira (Eds.), *Etnomatemática: papel, valor e significado*. São Paulo: Zouk. Furthermore, the §5 in my book *Travelling through education*, Sense Publishers, 2005, has been elaborated on the basis of the paper.

[48] My use of disposition can be compared to Bourdieu's concept of habitus:

> The habitus is the set of dispositions which incline agents to act and react in certain ways. The dispositions generate practices, perceptions and attitudes which are "regular" without being consciously co-ordinated or governed by any "rule". (Thompson, in the *Introduction* to Bourdieu, 1991: 12)

However, the notion of foreground does not play a vital role in Bourdieu's notion of habitus, which appears to be more related to 'background' in my interpretation. While background (and habitus) concerns experiences, foreground has to do with expectations, although certainly mediated by experiences. Nevertheless, I find Bourdieu's concept of habitus extremely interesting as part of the clarification of both the notions of foreground and background.

[49] Brentano's main work *Psychologie vom empirischen Standpunkte* is from 1874. An English translation is found in Brentano (1995). Searle (1983) analyses conceptual connections between action and intentionality.

[50] With reference to classic philosophic interpretations, meaning is assumed to be a referential property. The meaning of a concept is the set of objects or entities to which the concept refers. According to this interpretation, when we in mathematics education are able to expand the references of notions such as function, exponential function and metric space we provide extra meaning of such concepts.

REFERENCES

Alrø, H., & Skovsmose, O. (2002). *Dialogue and learning in mathematics education: Intention, reflection, critique*. Dordrecht: Kluwer.

D'Ambrosio, U. (2001). *Etnomatemática: elo entro as tradicões e a modernidade*. Belo Horizonte: Autêntica.

Bishop, A. (1990). Western mathematics: the secret weapon of cultural imperialism. *Race and class, 32*(2), 51-65.

Bopape, M. (2002). *Mathematics school based in-service training (SBINSET): A study of factors contributing towards success or failure of SBINSET in the South African school context*, unpublished doctoral dissertation (contact e-address: Bopapem@nu.ac.za). Aalborg, Department of Education and Learning, Aalborg University.

Bourdieu, P. (1991). *Language and symbolic power* (Thompson, J., ed. and *Introduction*). Cambridge: Polity.

Bourdieu, P. (1996). *The state nobility: Elite schools in the field of power*. Cambridge: Polity.

Brentano, F. (1995). *Psychology from an empirical standpoint* (2nd ed.). London: Routledge.

Castells, M. (1998). *The information age: Economy, society and culture* (vol. III: End of millennium). Oxford: Blackwell.

Civil, M., & Planas, N. (2004). Participation in the mathematics classroom: Does every student have a voice? *For the Learning of Mathematics, 24*(1), 7-12.

Ginsburg, H. (1997). The myth of the deprived child: New thoughts on poor children. In A. Powell & M. Frankenstein (Eds.), *Ethnomathematics: Challenging eurocentrism in mathematics education*. Albany: SUNY Press, 129-154.

Gorgorió, N., & Planas, N. (2000). Researching multicultural classes: A collaborative approach. In J. Matos & M. Santos (Eds.), *Proceedings of the 2nd International Mathematics Education and Society Conference*. Lisbon: Centro de Investigação em Educação da Faculdade de Ciências, Universidade de Lisboa, 265-274.

Gorgorió, N., & Planas, N. (2001). Teaching mathematics in multilingual classrooms. *Educational Studies in Mathematics, 47*(1), 7-33.

Khuzwayo, H. (1997). *[A report on] Mathematics Education in South Africa: a historical perspective from 1948-1994* (contact e-address: hbkhuzw@pan.uzulu.ac.za). Copenhagen, Department of Mathematics, Physics, Chemistry and Informatics, Royal Danish School of Educational Studies.

Khuzwayo, H. (2000). *Selected views and critical perspectives: An account of mathematics education in South Africa from 1948 to 1994*, unpublished doctoral dissertation (contact e-address: hbkhuzw@pan.uzulu.ac.za). Aalborg: Aalborg University.

Lins, R. (2001). The production of meaning for algebra: A perspective based on a theoretical model of semantic fields. In R. Sutherland, T. Rojano, A. Bell & R. Lins (Eds.), *Perspectives on school algebra*. Dordrecht: Kluwer, 37-60.

Penteado. M. (2001). Computer-based learning environments: Risks and uncertainties for teachers. *Ways of Knowing, 1*(2), 23-35.

Penteado, M., & Skovsmose, O. (2002). *Risks include possibilities, a report*. Copenhagen, Roskilde and Aalborg: Centre for Research in Learning Mathematics, Danish University of Education, Roskilde University and Aalborg University.

Planas, N., & Civil, M. (2002). Understanding interruptions in the mathematics classroom: Implications for equity. *Mathematics Education Research Journal, 14*(3), 169-189.

Searle, J. (1983). *Intentionality: An essay in the philosophy of mind*. Cambridge: Cambridge University Press.

Skovsmose, O. (1994). *Towards a philosophy of critical mathematics education*. Dordrecht: Kluwer.

Valero, P. (2002). *Reform, democracy, and mathematics education: towards a socio-political frame for understanding change in the organization of secondary school mathematics*, unpublished doctoral dissertation (contact e-address: paola@learning.aau.dk). Copenhagen, Department of Curriculum Research, The Danish University of Education.

Volmink, J. (1994). Mathematics by all. In S. Lerman (Ed.), *Cultural perspectives on the mathematics classroom*. Dordrecht: Kluwer, 51-68.

Wedege, T. (1999). To know – or not to know – mathematics, that is a question of context. *Educational Studies in Mathematics, 39*(1-3), 205-227.

Wittgenstein, L. (1953). *Philosophical investigations*. Oxford: Blackwell.

AFFILIATIONS

Ole Skovsmose
Department of Education and Learning,
Aalborg University

RENUKA VITHAL

THE "UNCIVILISED" SCIENTIST

Lessons from Ethnomathematics and Critical Mathematics Education for an African Scholarship[51]

"I am an uncivilised scientist". This is how a chemistry professor now engaging a doctoral study in science education described himself. It also captures the bewilderment and betrayal that many of the science graduates who enter education and social sciences studies experience when they come to address the history and foundations of their discipline and come to see that the so-called "scientific method" is but one of many different ways of knowing the world. This is often observed among science graduates who are becoming teachers and those who choose to research some educational dimension of their discipline. Several recent studies that offer in depth analyses of student reflections and experiences of programmes and curricula raise devastating critiques of science and related faculty curricula and pedagogies (Beecham, 2002; Pillay, 2003; Kathard, 2003). While much is being learnt about the outcomes and consequences of particular science programmes, the key challenge lies in feeding such research back into curricula to bring about curriculum reforms or transformations within university education.

The recent significant reshaping of the higher education landscape through mergers in post-apartheid South Africa has forced into the open deeper concerns about university curricula. Questions are being raised about the kinds of knowledge being produced, by whom, where, and in whose interests; and more specifically, questions about the kinds of researchers and professionals emerging from university and the quality of their knowledge, skills, values, attitudes and competence to function in a changing, unequal and diverse society. The merger of the University of Durban-Westville, a former historically disadvantaged university, with the University of Natal, a former historically advantaged university, to create a new University of KwaZulu-Natal has produced a unique window of opportunity to ask foundational questions and fundamentally rethink curricula to both redress our colonial and apartheid legacies and craft a new institutional identify for the future.

It is within this context that the new vision and mission of the university has put forward the idea to develop and promote an African scholarship. But what does the notion of an African scholarship actually mean or could mean for a new South African university? The attempt to engage different interpretations of African scholarship across the university in the process of rebuilding a new post-apartheid inclusive democratic institution, created a space for reflecting on recent

U. Gellert, E. Jablonka (eds.), Mathematisation – Demathematisation: Social, Philosophical and Educational Ramifications, 95–105. © 2007 Sense Publishers. All rights reserved.

developments in mathematics education which could provide a source and may be instructive for reconstructing university mathematics and science curricula for life in South Africa and for the world in the twenty-first century.

The last thirty years have seen significant growth of mathematics education research, theory and practice in new areas that advance debates and understandings of the historical, sociological, cultural, economic, philosophical and political aspects of teaching and learning mathematics which interrogate issues of class, race, gender, language and so on. In this paper I focus on two developments in mathematics education that may be considered to be a part of this growth, that of ethnomathematics and critical mathematics education, to exemplify what may be construed as elements of an African scholarship and explore how university mathematics and science education could be "humanised" or "civilised" and rescued from its cold harsh hard portrayal. The assumption being made here is that an African scholarship values an integrated connected knowledge (e.g. *Ingede: Journal of African Scholarship*) and these ways of knowing are explored and illuminated in and through fields of work such as ethnomathematics and critical mathematics education.

Ethnomathematics and critical mathematics education are part of a broad landscape that draws on a diverse range of disciplines to explore the social, cultural and political dimensions of mathematics and mathematics education that I have spent the better part of the last decade studying and integrating into teacher education programmes (Vithal, 2003). Arguably, they are also built into the new South African national school mathematics curricula where mathematics is defined as a "distinctly human activity practised by all cultures" and among its many purposes is to "empower them (learners) to make sense of society" (Department of Education, 2003: 9). This begs the question of what are the implications of the new school curriculum for students continuing with mathematics and science within higher education. This paper turns its gaze therefore to university curricula, to the education and training of scientists, and questions their programmes' resilience, (some might say) resistance to change and relevance to the world today. What kind of world must the production of an African scholarship face in countries like South Africa?

ETHNOMATHEMATICS

Ethnomathematics originated in the former colonies, in response to the eurocentricsm of mathematics and mathematics education and its consequences. Ubiratan D'Ambrosio, Brazilian mathematician, philosopher and educator, and founder of ethnomathematics both as a field of study and practice maps out the paradox that science faces:

> In the last 100 years, we have seen enormous advances in our knowledge of nature and in the development of new technologies. [...] And yet, this same century has shown us despicable human behaviour. Unprecedented means of mass destruction, of insecurity, new terrible diseases, unjustified famine, drug

abuse, and moral decay are matched only by an irreversible destruction of the environment. Much of this paradox has to do with an absence of reflections and considerations of values in academics, particularly in the scientific disciplines, both in research and in education. Most of the means to achieve these wonders and also these horrors of science have to do with advances in mathematics. (D'Ambrosio, 1994: 443)

In the development of ethnomathematics as a response to this paradox of science, D'Ambrosio emphasises the socio-cultural and socio-political basis of mathematics and its connectedness to context. A key argument that ethnomathematicians make is that mathematical knowledge is "embedded" within the knowledge systems of communities and cultures and requires a deliberate, careful and sensitive excavation and restoration. A vast array of studies identifying the mathematics of a wide range of different peoples, both past and present is now available even as questions of what exactly is the nature and value of this "embedded knowledge" and how is it to be known without reference to "academic", "Western" or "European" mathematics continue to be debated. The point to be made here is that much less attention has been paid to who should take up this task of mathematical archaeology – mathematicians or those within the communities and cultures whose knowledge is being explored; and what kind of university mathematics education, curriculum and pedagogy is needed to produce mathematicians or scientists who could or would undertake this?

This metaphor of the "embeddedness" of knowledge is also useful for exploring another aspect – for instance from the recent notions of "embedded journalists" being used in war situations as being co-opted and privileged with information as part of the mechanism of waging battle. Following this meaning we may ask what of *"embedded scientists"* whose products and labour is used in the service of wars and conflicts. This raises questions of how in the training and education of scientists is this accounted for. Since ethnomathematics recognises the value-laden nature of mathematical work, its inclusion in curricula must force at the very least questions about the historicity and values in mathematics and science teaching and learning in universities. This brings to the fore that scientists are in a sense always "embedded" in the society and the times in which they work and choose to engage particular problems and questions. Ethnomathematics explodes the myth of neutrality not only philosophically and sociologically but also pedagogically in the reality of the experienced mathematics and science curriculum.

Ethnomathematics, as a field of study and of practice has been developing over the last thirty years. It has a deliberate agenda to recognise the mathematics of different social and cultural groups in society both historically and in the present.

A broad range of writings is typically brought together in explaining ethnomathematics. Elsewhere, I have provided a detailed synthesis of the main contributions of ethnomathematics in terms of four broad strands (see Vithal, 1993a; Vithal, 1993b; and Vithal & Skovsmose, 1997). The first strand challenges the conventional history of mathematics [demonstrating that the mathematics we have today has its origins not only in Greece but has drawn on and builds on the intellectual work of many other peoples – Egypt and North Africa, the Middle East, China, India] (e.g. Joseph, 1991) and draws attention to the mathematical histories of cultures outside Europe and the Western world [ancient Incas, Mayans, Aborigine, Africans, etc]. The second strand focuses on the mathematics of traditional cultures, of indigenous people found in all parts of the world [in their cultural practices and artefacts: kinship relations, games, basket weaving, art, architecture and building constructions, etc have been mathematically analysed] (e.g. Ascher, 1991). The third strand elaborates the mathematics of different groups in societies such as carpenters (Millroy, 1992), child street vendors (Carraher, 1988); shoppers (Lave, 1988) and candy sellers (Saxe, 1990) [and numerous other studies on the work of fisherman, builders, dressmaking, gambling, dairy workers, etc]. The fourth strand explores the relationship between ethnomathematics and mathematics education (e.g. Pompeu, 1992, Vithal, 1993b). It is this last strand that has been my main interest and is also important for our purposes here. A main concern has been with the implications of ethnomathematics for school mathematics [and here it is being considered in relation to higher education in general and science faculties in particular].

The fundamental challenge put out by ethnomathematics to both mathematics and mathematics education is aptly captured in the title of a recent book "Ethnomathematics: Challenging Eurocentricism in mathematics education", edited by Arthur Powell and Marilyn Frankenstein (1997). In this volume of diverse writings drawing on anthropology, cognitive psychology, historical and feminist studies, mathematics and mathematics education, Frankenstein and Powell outline the main goal of ethnomathematics as follows: "The book challenges the particular ways in which Eurocentrism permeates mathematics education: that the academic mathematics taught in schools world-wide was created solely by European males and diffused to the Periphery; that mathematics knowledge exists outside of and unaffected by culture; and that only a narrow part of human activity is mathematical" (p. 2). Ethnomathematics has no doubt forced a re-examination of what constitutes and counts as mathematical knowledge, how it is and continues to be produced and legitimated, and who has been recognised for its production. It has also forced a re-consideration of theories and practices in mathematics education, and in particular, it has problematised the link between mathematics curricula and their relevance and appropriateness for different socio-cultural groups they are intended to serve. (Vithal, 2003: 16-17)

Although the term ethnomathematics does not rest easily in the South African context, and I have written a critique of its use (Vithal & Skovsmose, 1997), particularly given the corruption and racialisation of the concept of culture during apartheid, it has nevertheless developed a significant scholarship across the world in the indigenous mathematics found across societies and especially in the thinking and practices of marginalised peoples. There is no question that a substantial indigenous mathematical knowledge base exists and a significant scholarship has developed related to it, but what must be raised here is the challenge of why has it not entered into the mainstream of mathematics and science undergraduate and postgraduate programmes?

CRITICAL MATHEMATICS EDUCATION

Critical mathematics education has emerged inside Western highly technological societies as a reaction to the naïve optimism of industrialisation as a progressive force developed hand in hand with capitalism and its power structures (Vithal & Skovsmose, 1997). Both ethnomathematics and critical mathematics education are concerned with the role and function of mathematics in society, the queen of the sciences, as a threat to democracy and seek a deeper (perhaps less optimistic) understanding of its relation to progress and development, through its application in science and technology.

Ole Skovsmose, a Danish mathematics educator and philosopher of science and technology, addresses the link between democracy and mathematics through the concept of the "*formatting power of mathematics*". He argues that

> "mathematics produces new inventions in reality, not only in the sense that new insights may change interpretation, but also in the sense that mathematics colonises part of reality and reorders it" (Skovsmose, 1994: 42). A main concern for critical mathematics education has been with understanding the nature of this formatting power and developing the means by which all people can engage with the mathematically formatted nature of society. [...] Related to the formatting power of mathematics is a mathematical archaeology, the process of digging mathematics out and making explicit the actual use of mathematics hidden in social structures and routines
>
> There appears to be two parallel concerns at play. There is the concern to understand the relation between society and mathematics as a discipline on the one hand, and its relation to mathematics education on the other hand. In the former, one concern has been the ways in which mathematics is used to format society through all types of application. Society is increasingly mathematised. Mathematics as a discipline grows through processes internal to itself but it also grows through factors outside such as through applications, most obviously in science and technology but also to social or humanistic areas. The complexity of this relationship is demonstrated through the notions of real or social abstractions and thinking abstractions (see Keitel, 1993; Keitel, Kotzmann & Skovsmose, 1993; Skovsmose, 1994). As mathematics is applied to all sorts of structures, processes and organisations in society, these in turn change and require further mathematisation. The main idea is that an increasing amount of implicit mathematics exists in all facets of society and assumes a more sophisticated mathematically literate person to question how the applications are used within a democracy.

This is without doubt also an important concern of any 'new' or developing democracy, especially when the literacy and more so, mathemacy rates are very low. But an equally important concern in mathematics education, particularly in developing nations such as South Africa, has been to produce those people who can participate in the formatting power of mathematics – in creating the models, processes and structures by which society is becoming mathematically formatted. [This is typically the function of science and related faculties.] A critical mathematics education therefore must embody this two-fold aim, a concern for those participating and implicated in the formatting as well as those outside who need to react, since both have implication for the kinds of democratic competences and participation made possible in a democratic society. A mathematics education has to provide both the explicit knowledge and skills to actually use and apply mathematics, and to question and understand their implicit use (and misuse) in society. The existence of the first competence does not simply imply the second because the language of mathematics itself does not provide the means for criticising its application in society. This critical competence in being able to apply mathematics and also to evaluate its consequences according to social, ethical and political concerns must become a part of the mathematics education of both those who participate in the formatting power of mathematics including those in governance in a democracy, as well as those who have the position of mainly reacting to that power – the electorate. Mathematics is used in a multitude of ways in society – to predict, control, interpret, describe, and explain within a particular socio-political context. Mathematical solutions, especially to social problems, for example distributing social welfare, are given a particular value in society, often as neutral and objective solutions. A mathematics education must create opportunities for students not only to come to understand the implicit nature of mathematics in society but also the processes and means by which mathematical solutions come to be selected and given priority as well as the advantages and limitation of such choices. [...] A mathematics education for democracy is not therefore what is often referred to as functional or practical mathematics usually offered in numeracy programmes, nor is it assured through knowing more advanced mathematics. It is a far more complex competence which integrates a mathematical competence with a critical competence that includes social, political and ethical concerns. (Vithal, 2003: 8-9)

If the above argument is plausible then it must be asked where and how is this integrated competence to be produced in the education and training of scientists. The above points to producers, users and consumers of mathematical, scientific, and technological skills and knowledge and the different qualities that must be embodied in the education of these groups. There is yet another group outside this direct gaze of mathematics and mathematics education, but who nevertheless experience its consequences.

When placing both ethnomathematics and critical mathematics education in the globalisation debate then what may be observed is another paradox – a process of inclusion through information communication technology but also simultaneous exclusion by lack of access to such technology. It has been argued that globalisation unites as it divides. The important point for an African scholarship is that it produces new shifting fault lines. That is, an increasing ghettoisation goes hand in hand with globalisation – and creates what Skovsmose (2003) calls "disposable people" to which mathematics contributes in large measure both directly and indirectly not least through stratification and examinations. Castells (1998) presents this as the "Fourth World – the black holes of informational capitalism. These are the regions of society from which statistically speaking, there is no escape from the pain and destruction inflicted on the human condition of those who, in one way or another, enter these social landscapes." For Castells the rise of the fourth world which is found everywhere, both in wealthy and poor countries, though in different degrees, includes the poor, the homeless, the sick, the illiterate, the brutalised and criminalized; and as he argues, is inseparable from the rise of informational global capitalism. Mathematics science and technology are deeply implicated in producing and reproducing these different worlds, and hierarchies in relation to the development of knowledge societies. We need only look at the differential knowledge and skills that new students entering our university bring, given the enormous inequalities of schooling, to see how higher education participates in how it invites or excludes through its administrative and educational processes and procedures.

If this analysis is accepted then the education and training of scientists must be a preparation for acting in and against such a world. But by and large we do not see a groundswell of scientists protesting the use of their labour in projects against humanity. Recently the lack of academics engaging a public intellectualism in South Africa has be raised in the media and analysed largely from a political point of view. But from a different vantage point, that of *curriculum analysis*, what is made visible are the silences and gaps in the educational and research development programmes of scientists. That is the narrow technical focus in their specialist disciplines fails to provide the language and curriculum tools to engage a critique and contextualisation of their discipline. Nowhere in the education (e.g. assessments) of science graduates or academics or in the criteria for evaluating academic work is the translation or recontextualisation of disciplinary knowledge and skills into a public discourse made explicit or valued. This might also explain, in part, why for example the rank and file of scientists did not flood the media, or use the opportunity to enhance the public understanding of science and its rules and regulations of knowledge production when the HIV/AIDS debate opened in South Africa challenging the very foundations of science.

An African scholarship is needed to be developed but not only in opposition to a European or any other scholarship, rather it needs to be developed on its own terms in relation to the problems and paradoxes of the local and the diverse global. Notwithstanding complaints of curricula overload in higher education, questions must be asked about why particular selections exist, what purpose they serve and whose interests are being realised. Such questions serve to democratise, to civilise and humanise science – this is a specific quality that an African scholarship has the potential to bring to science and to the education and training of scientists. Such a task requires and assumes an appropriation of particular meanings and values related to what it means to live and work in Africa integrated into the ways of knowing and doing mathematics and science. It also implies the imperative to create and engage a serious and rigorous scholarship, developing relevant and appropriate criteria of quality within and for an African scholarship.

IMPLICATIONS

Curricula that seek transformation for the production of a new generation of African scholars, need to address some difficult questions and challenges raised through ethnomathematics and critical mathematics education. If an African scholarship is to be recognised and realised in teaching, learning and research in higher education then several important shifts will need to be made, some of which are outlined below.

First, opportunities need to be created for students to raise and address questions about where the knowledge they are expected to learn comes from - who produces it, in what socio-historical, cultural context, and in response to what political and economic imperatives. That is, to know *the histories and foundations of their discipline*, not only as a development begun in Europe but a more authentic investigation that recognises the contribution of all peoples – at the very least key moments of revolution that humanises and puts into a socio-political and cultural context why particular skills and knowledge turns where made. Engaging the philosophies and roots of their discipline must debunk and demystify the Euro-centrism so deeply entrenched in theories and practices of the inherited curricula of universities of the former colonies. This of course has major implications for who is to teach such ideas and the knowledge base and ideological orientations of academics. An African scholarship must at the very least require this engagement within existing programmes or in developing new ones if it is move beyond rhetoric.

The second is the recognition of the need *to work in the interface of the boundaries of disciplines*. A democratisation of mathematics and science goes beyond the acquisition of the technique and tools of a discipline. But a multi-, inter-, and especially a trans-disciplinarity requires particular structures and processes that allow for dialogue and the development of programmes that can produce, for example, philosophers, historians or sociologists of science, mathematics and technology. No doubt border crossings between faculties for such developments to occur are often difficult but the assumption is that an African scholarship requires and is given meaning by an open rather than bounded concept of knowing. An African scholarship that takes scholarship seriously will need to develop not only new domains but will also need to develop consonant criteria of quality and rigor that can work across and between disciplines if it is to preserve excellence and not escape into a relativism.

Third, if it is acknowledged that higher education produces the "formatters of society" – those who come to occupy positions of power in society – then there is need to think about how in the curricula of graduates and researchers opportunities are created to *develop social and cultural awareness and political and ethical responsibility.* So that those in power, as an integral part of their education, understand how their power and privilege is linked to the marginalised of society and demonstrate a concern for how their knowledge and power is enacted in the interests of the third and fourth world that dominate in a country like ours. The politics of knowledge is seldom understood and engaged in science programmes as foundational knowing that deepens and adds value to becoming more rounded scientists and researchers. Sometimes offered as electives or narrow ethics courses, research students and academics are unable to enter into a public discourse about the broader meanings and relevance of their disciplines. The failure to acquire tools for a critical engagement of their disciplines and research outside the academy limits their possibilities to advance their disciplines in broader society. Critical mathematics education, for example, addresses precisely this concern in the teaching and learning of mathematics.

Finally, an African scholarship presupposes an *integral connection to community.* It recognises and creates space for collective reflection and action rather than individual experts and flattens the hierarchy in which all participants are deemed having knowledge to deal with improving life conditions. This implies that group and project initiatives that value and take serious the knowledge and skills of communities, need to become part of the education and training of practitioners and researchers. The huge research drive in South Africa to develop and promote indigenous knowledge systems has seen community leaders and elders drawn on as key participants in research projects. Ethnomathematics, for instance, has developed an increasing body of work through such projects. But what has not been observed to the same extent is a parallel translation of the outcomes of such projects into the mainstream university curricula, especially in undergraduate programmes. For instance, bringing the leadership of communities into the university classroom to share knowledge and skills. Developing an African scholarship requires breaking down the division between different forms of knowing, especially formal knowing and other related forms.

CONCLUSION

The specificities of curricula transformation toward an African scholarship need to be developed through an ongoing dialogue and interplay of theory, policy, practice, research and context. Whatever form it may come to take, remaining in constant critical reflection about what an African scholarship could mean and come to be realised in the real lives of people who inhabit higher education, is crucial.

NOTES

[51] A shorter version of this paper presented at the Ingede African Scholarship Conference, University of KwaZulu Natal, 23-25 March 2004, also appears online in *Ingede: Journal of African Scholarship, 1*(1).

REFERENCES

Ascher, M. (1991). *Ethnomathematics: A multicultural view of mathematical ideas.* California: Brooks/Cole Publishing Company.

Beecham, R (2002). A failure of care: A story of a South African speech and hearing therapy student, unpublished doctoral dissertation, University of Durban-Westville, Durban.

Carraher, T. N. (1988). Street mathematics and school mathematics. In A. Borbas (Ed.) *Proceedings of the Twelfth International Conference for the Psychology of Mathematics Education* (vol. 1). Veszprem: Ferenc Gerzwein, O.O.K, 1-23.

Castells, M. (1998). *The information age: Economy, society and culture* (vol. III: End of millennium). Oxford: Blackwell.

D'Ambrosio, U. (1994). Cultural framing of mathematics teaching and learning. In R. Biehler, R. W. Scholz, R. Sträßer & B. Winkelmann (Eds.) *Didactics of mathematics as a scientific discipline.* Dordrecht: Kluwer, 443-455.

Department of Education (2003). *National curriculum statement grade 10-12 (general) mathematics.* Pretoria: Department of Education.

Joseph, G. G. (1991). *The crest of the peacock: Non-European roots of mathematics*, London: Tauris.

Kathard, H. (2003). Lifehistories of people who stutter: On becoming someone, unpublished doctoral dissertation, University of Durban-Westville, Durban.

Keitel, C. (1993). Implicit mathematical models in social practice and explicit mathematics teaching by applications. In J. de Lange, C. Keitel, I. Huntley & M. Niss (Eds.) *Innovation in maths education by modelling and applications.* Chichester: Ellis Horwood, 19-30.

Keitel, C., Kotzmann, E. & Skovsmose, O. (1993). Beyond the tunnel vision: Analysing the relationship between mathematics, society and technology. In C. Keitel & K. Ruthven (Eds.) *Learning from computers: Mathematics education and technology.* Berlin: Springer, 243-279.

Lave, J. (1988). *Cognition in practice: Mind, mathematics and culture in everyday life.* Cambridge: Cambridge University Press.

Millroy, W. (1992). *An ethnographic study of the mathematical ideas of a group of carpenters* (JRME Monograph No. 5). Virginia: National Council of Teachers of Mathematics.

Pillay, M (2003). (Re)positioning the powerful expert and the sick person: The case of communication pathology, unpublished doctoral dissertation, University of Durban-Westville, Durban.

Pompeu, G. (1992). Bringing ethnomathematics into the mathematics curriculum, unpublished doctoral dissertation, University of Cambridge, Cambridge.

Powell, A. & Frankenstein, M. (1997). *Ethnomathematics: Challenging eurocentrism in mathematics education.* New York: State University of New York.

Saxe, G. B. (1990). *Culture and cognitive development: Studies in mathematical understanding.* Hillsdale: Lawrence Erlbaum.

Skovsmose, O. (1994). *Toward a critical philosophy of mathematics education.* Dordrecht: Kluwer.

Skovsmose, O. (2003). Ghettorising and globalisation: A challenge for mathematics education, publication no. 39, Centre for Research in Learning Mathematics, Denmark.

Vithal, R. (1993a). Ethnomathematics: Some questions for curriculum reconstruction. In C. Julie, D. Angelis & Z. Davis (Eds.) *Political dimensions of mathematics education 2: Curriculum reconstruction for society in transition* (conference proceedings). Broederstroom,154-167.

Vithal, R. (1993b). Ethnomathematics: Research directions and some implications for curriculum. In V. Reddy (Ed.) *Proceedings of the First Annual Conference of the Southern African Association for Research into Science and Mathematics Education.* Rhodes University, 334-351.

Vithal, R. (2003). *In search of a pedagogy of conflict and dialogue for mathematics education.* Dordrecht: Kluwer.

Vithal, R., & Skovsmose, O. (1997). The end of innocence: A critique of "ethnomathematics". *Educational Studies in Mathematics, 34,* 131-157.

AFFILIATIONS

Renuka Vithal
Faculty of Education,
University of KwaZulu-Natal

CLIFF MALCOLM

DIVIDING THE KINGDOM

WAYS OF THINKING

I was brought up on an Australian farm, not far from Melbourne. In the latter part of my Grade 11 year at high school, the Education Department sent in a team of vocational counsellors to help us plan our futures. The counsellors were kindly people who administered questionnaires and tests, and, next day, met with us to discuss the results. I had written on my forms that I hoped to go to university and study physics. The counsellors queried my choice: my aptitudes and interests clustered around poetry and stories, drawing, acting, singing.... Hardly the stuff of physics. Or is it?

I was unsettled by the counsellors' report and advice. Did their scales and scores offer a more accurate account of 'me' than I had through reflections and conversations I had engaged in over years with teachers, family and friends? Were the counsellors' mathematical analyses and representations 'truer' than my qualitative, intuitive ones? Were science and humanities so separate that if I chose one I would close out of the other, not only in my activities, but in the friends I would have? Was my self to be divided, set against itself?

I mulled again over these problems. Certainly, I loved the romance of life on the farm – the cold in my face as my horse galloped across morning frost, the bellow of new-born calves, the carolling of magpies high in the gum trees, the freedom of swimming naked in silent dams. But I enjoyed the other side of farm life as much: machines, tractors and tools, do-it-yourself solutions to plumbing, fencing and building, theories of silage and hay, detailed plans for cropping and breeding. And I knew how much I loved physics – Newton's laws, standing waves, harmonics on strings and cymbals, and the wondrous domains of atoms. Add mathematics, not only as a tool for solving problems and making predictions, but as a comment on 'reality'. How was it that the exponent in Coulomb's Law for interacting electric charges was precisely two? If light travelled from A to B by the path requiring least time, as Fermat argued, how did it make the calculations? I wondered whether mathematics in vocational counselling was the same as mathematics in physics.

As it turned out, I chose a career in physics and science education, and I'm happy with that. But I'll never know where the other path might have led, and I don't really know what the major influences were on my decision at that time. Not only did I have to choose in those weeks a career suited to one or another way of thinking (science-arts, mathematical-conceptual, rational-intuitive), I had to decide, *as part of the decision*, which mode of thinking to use. Is it possible that I chose science as a reaction against the counsellors' instruments, numbers and assurances?

U. Gellert, E. Jablonka (eds.), Mathematisation – Demathematisation: Social, Philosophical and Educational Ramifications, 107–122. © 2007 Sense Publishers. All rights reserved.

Did my decision hinge less on arguments and data than how the case was presented? Was I concerned less about what made one option acceptable than what made the other rejectable?

Writers of advertising and producers of television news know these things. They lean sometimes on the authority of scientific arguments, evidence and graphs, sometimes on human stories, images and lighting, music and rhythm. In any case, they know that some presenters' personalities, voices and styles will turn people off. Similarly, newspaper journalists know the importance of the first paragraphs of an article. A message is not only about content; it is about audience and the audience's interests and preferences. Science teachers and texts take a different approach, focussed on what they want to teach much more than what and how students want to learn. They emphasise the formal, impersonal, rational, mathematical aspects of science not because they think students relish them, but because that's the picture of science they want to present.

Teachers and science texts have advantages over journalists and advertising writers when it comes to audience. Students in the classroom are captive. They have contracted to learn, just as teachers have to 'teach'; compliance is built in. Second, teachers can hand out rewards and punishments (including assessments and promotions from grade to grade). Third, the classroom agenda and processes are, at least to some extent, participative and negotiable. Fourth, teachers and students stay together over time, enabling them to build trust and respect, incorporate variety, and repair errors. In spite of this, there is plenty of evidence that many students are not deeply engaged in their schooling. For example, Barber (1996) reported a large scale British study that indicated 20-30% of high school students were bored, lacked motivation, and were "disappointed" in their school experience, 10-15 % were openly hostile, "disaffected" with school, and 5-10% were frequent truants. In other words, half the students were somewhat disconnected from school. And now, as in the late 1960s, students are opting out of physical sciences in senior high school and post-compulsory education (e.g., Sjøberg, 2002).

My purpose in this chapter is to explore what makes science, as a way of knowing, acceptable or rejectable for students. I take as my starting point Social Judgement Theory (Sherif, Sherif, & Nebergall, 1965), and ways in which people use cognitive, attributional and heuristic thinking (Chaiken, Liberman, & Eagly, 1989) when making decisions. I link these ways of thinking to epistemological choices of intuitive and rational thinking and the 'two cultures' (science-maths and arts-humanities), and then to students' 'learning styles', using especially Herrmann's (1995) theories. Finally, I consider the ramifications for curriculum and teaching in science.

SOCIAL JUDGEMENT THEORY

In the 1960s, Sherif *et al.* (1965) wanted to understand what makes a particular message acceptable to an individual or group; what frames the judgements people make about that message. They called the theory Social Judgement and

Involvement, pointing to the importance of social factors and ego-involvement in the judging. They proposed that individuals sort incoming information, as it comes in, into a latitude (zone) of acceptance, a latitude of no-commitment, or a latitude of rejection. At the centre of the latitude of acceptance is an anchor position – a set of values and beliefs that the individual holds, and to which the message relates. Acceptance of a message is largely about the 'distance' between the anchor position and the position in the message: it is position that is critical, more than information. Individuals who are open to modifying their position in response to a message have large latitudes of acceptance for it, and small latitudes of rejection. However, their judgement depends on relevance as well as position: if they see the message as remote from their interests, they are likely to place it in the latitude of no-commitment. They don't care. If they think it is highly relevant and likely to reinforce and extend their existing beliefs, or enhance their well-being, their views of themselves, and their relationships with people who are important to them, they place it in their latitude of acceptance. If they think it is relevant but contrary to their interests, they place it in their latitude of rejection. The judgement is quite self-centred. So, for example, individuals might look for aspects of the message that they like or don't like and judge it according to these bits rather than critique the message as a whole.

Sherif *et al.* (1965) argued that people make judgements about an idea even as it is coming in, often more on bases of intuition and extrapolation than systematic analysis. In other words, no sooner has the message begun than the judging begins, and the judging continues as the message continues. First impressions might give way as the argument proceeds, or they might lock in. At any stage, judgements are based not only (or even especially) on the content of the message, but also the perceived values, authority and attractiveness of the speaker, the ways other people in the group view the message, current fashions and trends, and so on. Thus, while the criteria are essentially personal and ego-centred, social and cultural influences are part of that self-interest, arising directly through the relationships between presenter, audience and individuals, and indirectly through influences of groups and cultures on individuals.

I want to use Sherif *et al.*'s (1965) ideas in two ways. First, I want to consider students' ways of thinking (perhaps rational-empirical, perhaps not) as anchor positions. If the students' anchor position is not rational-empirical but the classroom message is, does that immediately place the science in the latitude of rejection? Second, I want to explore the ways of thinking students employ as they process a classroom message in relation to their anchor positions: does it help the 'non-science' student's engagement if the science message is presented using 'non-science' modes of thinking and judging?

COGNITIVE, ATTRIBUTIONAL AND HEURISTIC THINKING

Researchers such as Sherif *et al.* (1965) and Chaiken, Liberman and Eagly (1989) consider that the judgements people make are reasoned, in that individuals can articulate their thinking and justify their choices. One reasoning strategy available, the cognitive response, involves systematic, rational analysis of the content and structure of the message. Science teachers and schools are well used to this strategy, with its concerns for purpose, position, claims, evidence, logic and coherence in the message. In this case, judgement is driven mostly by data – data presented, and data from other experiences. A second strategy, the attributional response, is also systematic and analytical, but more theory-driven: it seeks to attribute causes and understand why the message has the content and form it has. Thus its concern is as much for subtext as text: What are the presenter's values? Where is she coming from? Why is she taking this position? Is her presentation a fair reflection of reality? What is she reacting against? What and who have influenced her? The attributional response is tilted to humanities approaches to critique, the cognitive response to science approaches.

A third strategy, heuristic thinking, is not systematic in the ways that cognitive and attributional strategies are. It is a kind of short-cut, responding to cues in a presentation and interpreting them through a small number of essentially simple rules. The rules might be based in stereotypes, prejudices, values, guiding principles, generalisations or even habits. Heuristic thinking can be a default strategy (when the individual does not wish to do the work of systematic thinking) or a conscious choice (when the content of a message is too far outside the individual's knowledge to permit systematic analysis). The cues can arise from the presenter, the message, or people in the immediate audience (or beyond it). For example, the presenter's attractiveness (personality, physical appearance and dress, values, background, trustworthiness, cooperativeness, preparedness to recognize and praise others) and authority are important. Two or three cues from the presenter can trigger attributions that result in rejection of the entire message. Cues arising from social influences include the judgements that others in the group are making, the ways individuals wish to be seen by and relate to the group, individuals' wishes to be seen as internally consistent, a general mood that the current opportunity should not be missed, and a sense of responsibility to 'give back' to the presenter and others who have already given. For cues arising from the content and structure of the message itself, such as its level of formality, its uses of data, mathematics, stories and pictures, I will draw from educational research into learning styles (below).

There is plenty of room for 'error' in the judgements people make! They face myriad pieces of information – arising from the message, the messenger, the environment, themselves, history – that might bear on the judgement. Perception and perceptual filters enable simplification, as individuals choose a focus, select information, and apply particular perspectives and values to that information. The process is highly intuitive and iterative, as perception chooses focus, and focus provides cues and data. This is so even when individuals choose their perceptual screens consciously (for example, by opting for an attributional analysis in terms of power). Further, because the message is usually judged from an ego-centric framework, Sherif et al. (1965) argue, individuals often unconsciously distort the message, making it more like their anchor position if they decide it is in their latitude of acceptance (assimilation), and less like their anchor position if they decide is in their latitude of rejection (contrast).

It is not clear how much individuals move back and forth between heuristic and systematic modes, left to their own devices. A presenter can guide the choice to some extent, for example, by starting with stories and images and then moving to more formal arguments. However, attempts to guide individuals' ways of thinking can backfire, if the imposed ways are too distant from individuals' anchor positions.

The anchor positions I am interested in are students' anchor positions in relation to science, and the distance (or not) between their positions and the ones presented in the science classroom. In general, an anchor position cannot be defined tightly, because it depends on the context, the messenger, and personal-social dynamics of interaction as well as the message. Its defining characteristic is that the individual chooses to identify with it and defend it – whether it centres on a worldview, a value, a principal or a belief. For example, students might place a claim that "science knowledge is tested, reliable and hence superior to other knowledge" in their latitude of acceptance during a study of machines, but in their latitude of rejection during a study of the origins of life. Or they might place it in their latitude of rejection for all contexts, rejecting *any* claim, assumption or implication that science knowledge is superior to other knowledge. This can happen in Western as well as non-Western cultures. Whether students opt to reject the claim of science as superior knowledge as a result of cognitive analysis, attributional analysis or heuristic thinking, their judgement can alert them to spot the same claim in other settings, and perhaps treat it as a cue to reject the science. If this happens, the 'main game' of science education is lost to a 'side play'; whether or not science is superior knowledge is not what science education is about.

INTUITION AND REASON

Students can take a position different from their science teachers over the relative merits of intuition and reason as ways of knowing. My Australian Oxford dictionary calls intuition 'immediate apprehension by mind, without reasoning'. This is consistent with early ideas of Pythagoras and Plato, for whom intuition was a kind of direct knowing, from mind to mind, mind to Truth, inner life to the cosmos. Thus intuition, by definition, is epistemologically different from reason: it happens 'without reasoning'; it is immediate, holistic, aesthetic, revelatory, inspired; it is feeling, knowing and believing all in one; it can go where reason cannot. All of us have experienced intuitive knowing, at least on occasions. In some people it is well developed.

While intuition is epistemologically different from reason, in practice intuition and reason work together in constant interplay. Consider, for example, intuition as part of our awareness of ourselves and things around us: children intuitively know their mothers' moods, teachers intuitively know their students' thinking, farmers intuitively know the weather, footballers intuitively know where the ball will go; I intuitively know whether I like someone I've just met. However, reason, background knowledge and information processing are part of this: teachers can be highly attuned to clues in students' body language, and well armed with knowledge of what students in the past have thought. They analyse and reflect, study and practice – honing their awareness, observations and knowledge – and hence sharpen their intuition. The 'direct knowledge' they feel might *be* the case (as background knowledge and keener awareness set up their intuitions), or *seem to be* the case (because they skip through observation and processing quickly and unconsciously, or bypass processing by calling on memories and habits). Either way, perception is critical: of all the data available teachers choose (consciously or not) to look for some clues, not others. Perception, intuition, observation and reason are not easily disentangled.

At a second level, intuitive knowledge is not so much findings as hypotheses to be tested: intuition produces an idea, and reason seeks to test or develop it. A teacher might feel intuitively that a particular student believes light reaches out from the eye, treat the intuition as an hypothesis, and check with the student; a painter might have an idea to represent the relationship between God and Man by having them reach out to each other, not quite touching, and then experiment with sketches to see if the idea is worth pursuing. The teacher's hypothesis might or might not prove fruitful; the painter's idea might or might not 'work'. Intuition now is about creativity, imagination, inspiration, revelation. It links to reason, but it is not reason. As Einstein observed, "I never came upon any of my discoveries through the process of rational thinking". His feeling of "coming upon" his discoveries is echoed by the novelist Danielle Steel (2002: 29):

> Where do ideas come from? I don't know... The story comes from somewhere and seems to flow through me until it's finished...

Einstein and Steel both point also to the importance of knowledge, analysis and hard work. Intuition and reason work together, interacting with each other. The 18th Century poet Wordsworth (quoted in Pooley, Anderson, Farmer, & Thornton, 1968) puts it nicely:

> Poems to which any value can be attached were never produced on any variety of subjects but by a man who... had also thought long and deeply.

At a third level, intuitive knowing is part of mystical experience: direct connection of the self to others and nature, revealing knowledge and truths. We know it as the "Ah-ha" experience, Archimedes' cry of "Eureka", and more. The knowledge might be able to be tested through reason, but it might not; it might not even be comprehensible by reason. Intuition is its own way of knowing. World religions have developed practices to enhance intuition, such as meditation, yoga, prayer, music and chant. But the mystical experience, whether an insight into unity or fragmentation, connectedness or alienation, joy or terror, beauty or ugliness, is not exclusive to religion: it is part of walking in the bush, attending a concert, working in a refugee camp, falling in love. People seek out mystical experiences in the things they do, in art, music, stories, poems, theatre, films, architecture. Mystical experiences are at the centre of cultural life. And they are at the centre of science – in the creation of hypotheses and data, the design of investigations and the functioning of perception.

Of course, hypotheses, inner convictions, dreams and brightly coloured imaginings can be wrong, perhaps deeply. Long lists can be assembled of false prophets, leaders whose inspirations had horrible consequences, mystics whose revelations were contradictory, and people in a particular situation who see the situation quite differently. In science, alongside amazing achievements, there has been an abundance of false starts, blind alleys, dashed hopes and terrible consequences. In the arts too, some pieces find greatness but most do not.

One way forward, as in the European Enlightenment, is to regard intuition as a process of knowledge generation, and ensure that the knowledge is tested through reason. This makes reason the final arbiter of the worth of the knowledge, and in that sense superior to intuition. It also gives special place to mathematics, as a model of reasoning and as an elegant and abstract representation of Truth. However, reason without intuition has little to work with and no obvious way of finding direction. Further, knowledge is distorted by requiring that only those aspects testable by reason are worth following through. An alternative, common in the arts and public life, is to celebrate intuition as well as reason, and set up systems of public critique that are aesthetic as well as analytical. Such systems already operate, through reviews, peer evaluations, commentaries and participation in the public media.

School science, traditionally, has taken the Enlightenment position, promoting and demonstrating reason, empiricism, objectivity and mathematical logic as the key characteristics of science, and generally placing the testing of ideas above the production of ideas. School science talks much about controlling variables, making measurements and drawing conclusions, but little about creativity and intuition. Consistent with this, it treats scientific laws and theories as discoveries more than inventions. Some science, such as Snell's law of refraction, might have been discovered, but most was invented. Newton had the imagination to ask, "What would happen if we said objects only change their motion as the result of an external force?" And then he followed through. Gravity and friction had to be considered as forces, and so did surface reaction. The Action-Reaction law became necessary. Further, Newton had the audacity to define three concepts (force, inertial mass, and an inertial frame of reference) in a single equation: $F = ma$. Not only did his system 'work', but it yielded the Law of Gravitation in a simple, elegant form. Newton did not discover the concept of force: he invented it, in ways that distinguished it from energy, momentum and pressure, and created a complete system. This discussion is seldom conducted in school science classes. Neither does school science discuss in any detail the ways in which Newton's intuition was sharpened by the spirit of the Renaissance, or the work of scientists such as Kepler, Galileo and Descartes – perhaps especially their conceptions of an inertia law and their struggles with action-at-a-distance. As Wordsworth observed, a poem might be written quickly, but no poem of worth was ever produced but by a person who had thought long and deeply.

Much more could be done in science teaching to acknowledge and celebrate intuition as well as reason in science, helping students build their intuitive response to science-related situations. This would present a fairer view of science, and it would be attractive to students who enjoy and encourage intuitive knowledge, who like to imagine and like to 'connect'. This could easily go beyond opportunities to create hypotheses and invent solution strategies, to production of plays, stories, games and simulations. The stories could be about the people of science, raising human and social issues, or about scientific perspectives of particular experiences. For example, students might have walked on a beach, feeling its immensity in something the way Lord Byron did *(Childe Harold)*:

Roll on, thou deep and dark blue Ocean – roll!
Ten thousand fleets sweep over thee in vain;
Man marks the earth with ruin – his control
Stops with the shore…

We could help them see other stories too, no less awesome: the sea as a teeming, tumbling mass of water molecules, held by gravity and intermolecular forces, a chemical medium for plants and animals and a physical medium for energy flows, tides and currents; waves and water bound to each other, wind and sun. One sea. A particular molecule, that one, might shortly be part of a cloud, a tree in the Amazon jungle, a child in the African desert, a river in China. And the molecule is almost entirely space! Substance without substance. A vacuum that is nothing and yet has electrical and magnetic properties, and can, with coaxing, yield mass. Too readily we take the beauty and mystery out of science. In doing so, we lose students who decide consequently that science is not for them.

MATHEMATICAL AND CONCEPTUAL SCIENCE

From Pythagoras and Plato, to Descartes and Newton, to Feynman and Hawking, philosophers and scientists have enjoyed the idea of mathematics as a rational, pure domain, separate from and transcending natural phenomena, and cherished the dream that stable and knowable laws of nature not only exist but can be expressed elegantly in mathematical terms. Certainly, examples abound of scientific laws and theories that have simple mathematical expression. Examples abound too of scientific concepts and theories in which mathematical representation is the best we have, because mental pictures and analogues tell only part of the story. Mathematics thus in one sense stands above science, and in another is its servant and partner. The extent of integration of the two depends on the science: in physics and cosmology it is high; in biology and psychology it is low. Sciences at one end of the scale are often called 'hard', at the other end 'soft'. The connotations of value are intended. (This was what bothered me, during my Grade 11, when my vocational counsellors offered me scores and scales concerning my interests: their 'hard' numbers were presumed to be more rigorous than my 'soft' analyses).

There is no question that school students, as they grow older, develop clear anchor positions in relation to mathematics and their abilities in mathematics. Their positions feed their response to science. Students whose position is "I'm no good at maths" have three options if they want to study science: choose a softer science, seek qualitative, conceptual approaches, or develop their mathematical abilities. For teachers in subjects like Physics it is no simple matter to present 'conceptual science' that avoids mathematical reasoning and representation: the mathematics is there, even if not formally so. Science curriculum designers and teachers understand these issues. It is part of a teacher's life to invent and trade in analogies, metaphors, games, models and simulations that provide alternatives – or supplements – to mathematical arguments. Understanding that the floor exerts a force on me is helped by visualising distortion of the floor. Thinking of mass as an amount of substance, not merely 'what is measured on a beam balance' is helpful.

However, there is another side to this. Teachers often retreat into mathematics, especially algorithms and formulas, instead of developing qualitative under-standing. "Given u, a and t, this is how you find v; In a circuit, given V and R, this is how you find I. Questions like this will be on the exam." Often this strategy is promoted more for students whose mathematics is weak than for those whose mathematics is strong. Rigour in understanding gives way to illusions of rigour, tied to mathematical representation and manipulation.

SCIENCE AND THE ARTS: TWO CULTURES

Historically, the counter to the European Enlightenment's emphasis on reason, mathematical thinking and Truth was Romanticism, which placed imagination, emotion, participation and free-will at centre stage and sought different truths. Romanticism found expression not in sciences but in the arts, through people such as Rousseau, Blake, Beethoven, Goethe, Wordsworth, Coleridge, Dostoevsky, Nietzsche. Science/maths and humanities/arts separated, as 'two cultures' (Tarnas, 1993: 366-378; Snow, 1959).

Eventually the Enlightenment and Romantic paths both ran into difficulties (Tarnas, 1993: 355-394). Within science, theories from Quantum theory to Jungian psychology to Chaos theory questioned the separability of subject and object and pointed to interconnection in nature. It became doubtful that humans could ever accurately describe and explain 'reality' – even in the limited realm of the material world. Meanwhile, the socio-political and environmental violence, destruction and fragmentation that characterized the 20th Century, in spite of immense improvements in health and material well-being, made the Romantic project seem quixotic, and called instead for 'realism' in describing contemporary life. The Enlightenment dream of Truth and the Romantic dream of Unity were both in trouble. Post-modernism had to deal with this.

For schools, philosophical developments over the last century have had little effect. The Enlightenment/Romantic division and its two cultures remain. Students now, as when I was in Grade 11, choose between science and humanities (or, more accurately, science and non-science) in order to fit their abilities, school timetables and university pre-requisites. Schools continue to present a narrow, discipline-centred, positivist version of science, and science continues to present itself as superior knowledge. The words *enlightened* and *romantic* have undertones that persist.

There is a further issue, critical in a post-modern world. The idea of two cultures is a Western idea, and arose in a Western context. But the world is bigger than Western, and there are many more cultures than two, not only on a world scale, but in any one country. In African philosophy, as in Buddhism, Taoism, Hinduism and many other non-Western systems of thought, there is no sharp distinction between intuition and reason, subject and object, arts and science. Yet science is taught in much the same ways, with much the same content, from much the same (Enlightenment) position, in all countries, to all cultures and all cultural mixes.

From a non-Western point of view, students have to ask whether it is a good idea for them to compartmentalise science in the Western way, and indeed why such a choice is even required. The African scholar Mpahhele (2002: 17) explains:

Whether we like it or not, we the subject race are – like X – carriers of a disinherited imagination. This is what I mean: to think of standards as good because they are established by the white man and his traditions. 'Universal excellence' refers to the white world and pretends that we are included simply because we, the people of colour, also inhabit this planet. When we adopt standards alien to our own considered ideals, we inevitably end up with X's half-baked concepts and half-truths....

As science educators, we have to ask whether a sense of the superiority of science (over the humanities and over other cultures) and the system that gave rise to such a conclusion is what we intended, whether we intend it now. With science a compulsory subject in most countries of the world, we have to ask whether our intention is to include all students, or to exclude some groups in favour of others.

FOUR WAYS OF THINKING, ONE BRAIN

The idea of two cultures, with their different ways of knowing, resonates with theories of left-brain/right-brain thinking, in which the left-brain represents rational-analytical thinking, and the right-brain creative-associative thinking. Herrmann (1995) extended this to "four quadrant, whole brain" theory. He imagined the brain hemispheres divided into front and back segments, as shown in Fig 1.

The left-brain encompasses rational thinking (front quadrant, A) and ordered-thinking (rear quadrant, B), and the right-brain encompasses feelings-thinking (rear quadrant, C) and creative thinking (front quadrant, D). The theory has some support from studies of the physiology of the brain but does not depend on them: its empirical basis is in studies of people's behaviours and preferences and the stability of their preferences as personal traits.

Herrmann's (1995) theory is helpful for three reasons. First, it offers a classification of individual learning styles and preferences that is well researched, resonates with teachers' knowledge of their students, and is directly and easily applicable in curriculum design. Second, it fits well with Social Judgement Theory, because each of its four kinds of thinking contains a clear position, a value, a preference, which can operate as an anchor position. An individual with a preference for creative thinking (D) is likely to reject science classes that are forever tightly prescribed with detailed instructions (B). Third, while the research shows that individuals generally have a preference for one or another way of thinking, they nevertheless use all of them, depending on what is required: creative thinkers (D) might abhor keeping records (B), but do so because it is necessary; rational-analytical thinkers (A) might shy away from expressing feelings in a classroom discussion (C), but be open about them in more intimate situations.

Fig 1. Four quadrants, one brain: four ways of thinking (adapted from Herrmann, 1995)

A: Rational thinking:	D: Creative thinking:
The power and logic of theories and generalizations; use of data and evidence. (scientists, engineers)	Creating, relating and representing ideas in new ways; seeing the big picture. (novelists, cartoonists)
B: Ordered thinking:	C: Feelings thinking:
Concerned for details and order; enjoying rules, set procedures, routines. (accountants)	Concerned for feelings, emotions, other people and lives; intuition, empathy (nurses, social workers)

This use of all four ways of thinking means that the kingdom, as a metaphor for the individual, is *not* divided in the ways that the two cultures suggest. Individuals follow their special talents, and use different ways of thinking at different moments, in different contexts, but they use all four modes, and use them to support each other. Thus Herrmann (1995) argues for the development of all modes, to open up flexibility and possibilities as part of solving problems, building relationships, thinking creatively, and getting things done. In terms of Social Judgement Theory too, students should have opportunities to use all four ways of thinking. Similarly, the classroom need *not* be divided in the ways that the two cultures suggest. Different students using different modes of thinking can work together to overall advantage. Science education does not need to define science in a way that separates the student population into those who can do science and those who cannot.

Herrmann (1995) developed inventories that measure individuals' relative preferences (or scores) for the various modes. One person's profile might be 'square' (the same score in each mode), whereas another's might be highly skewed. Applied to vocational groups, typical profiles generally fit the stereotypes. Engineers and scientists cluster around rational-logical thinking (A), and are generally weak in empathetic thinking (C). Conversely, nurses prefer intuitive, interpersonal thinking (C) over rational-logical thinking (A). Accountants tend to favour structured tasks, and organised, detailed information (B). Cartoonists and writers enjoy 'big picture' thinking, juxtaposition of ideas, visual approaches and 'what if…' questions (D). Teachers, overall, tend to square profiles, but subgroups (eg science teachers) can exhibit particular preferences.

At the University of Cape Town, Horak and du Toit (2003) used Herrmann's

(1995) theories and instruments to describe students and staff, texts, lectures and assessment in the university's civil engineering programmes. Curricula, teaching and assessment were almost entirely weighted to rational-thinking and ordered-thinking, with almost no attention to feelings-thinking and creative-thinking. Male students tended to be rational-thinkers, with female students stronger in feelings-thinking. Lecturers (male and female) favoured the rational-thinking and creative-thinking quadrants. However, in data from 420 science students from rural schools who were part of an academic enrichment programme at the University of Durban-Westville, I found no significant gender differences in, for example, students' preferences for "learning that starts from our lives and feelings", "working through theories logically, thinking about evidence and argument", and "learning from activities that can be done in different ways". The students liked to use a wide variety of learning strategies, and their preferences could be grouped into a small number of learning styles, but their preferences did not correlate with gender (Malcolm, 2004).

The critical point is that, in classes in science and engineering, there are students whose preferred thinking modes are in all quadrants, not just A. Further, science, except in the essentialised, distorted version presented in schools, uses all four thinking modes. In situations such as Horak and du Toit (2003) report, curriculum and teaching suits only some students, addresses only some dimensions of science, and, through selection processes, is self-perpetuating. Everyone loses – students and science. This is especially so when universities are struggling to attract and hold science students, and scientists and engineers are expected to be creative and well attuned to social values as part of their work. This was the situation that motivated Horak and du Toit's (2003) study.

BUILDING THE KINGDOM

In 1982, Posner, Strike, Hewson and Gertzog (1982), using a constructivist learning framework, proposed that students applied criteria of "intelligible, plausible and fruitful" when judging new knowledge in relation to their existing ideas. These criteria quickly became a mantra for curriculum design and teaching. But they are only part of the story. They centre on rational-analytical thinking (quadrant A) and cognitive response. They take little account of attributional analyses that students might make, or heuristic thinking that is probably common. They give little attention to emotional and social criteria, students' values and belief-systems, or non-Western and non-science cultures. In short, they start from science and assume either that students want to think that way, or 'should' think that way in a science class.

Social Judgement Theory invites us to consider not just what makes as idea acceptable, but what makes it rejectable. We are used to the idea in science that no amount of supportive data can prove an hypothesis, but a small amount of contrary data is enough to disprove it. Sherif et al.'s (1965) latitude of rejection is reminiscent of this: for the student, there are many more cues and clues that can result in rejection of an idea or non-commitment than accepting it. From Social

Judgement Theory, non-commitment is mostly about indifference, lack of involvement. In such a situation, examinations and social factors are the only reasons for students to get involved. Students 'play the game' of doing what is necessary to pass (or even do well) on the examinations, without concerns for the science. Following Larson (1995), Aikenhead (1996) has dubbed 'what is necessary' as Fatima's Rules. When students do feel personally involved in the science, they are likely to place it either in their latitudes of acceptance (in which case learning can occur) or their latitudes of rejection. If they reject it, they still have the option of playing Fatima's rules (Aikenhead, 2004): passing exams need not mean they believe or understand what they wrote. At a deeper level, as I noted earlier, it is possible for students to reject the whole of science on the basis of a message that might have been avoided or handled differently, such as when science is presented as superior knowledge, or when the school (or society) declares entire groups incapable of science because they don't think the right way. Less dramatically, students might place science in their latitudes of acceptance for some topics, but their latitudes of rejection for others. For example, African students might accept an explanation of rain-making in terms of the water cycle as interesting though limited; they might reject it outright if they feel it denies roles also for rain-makers, ritual dances and spirits (Malcolm & Keogh, 2004).

My concern here is not for some kind of political correctness, whereby teachers avoid giving offence, or to put teachers on tenterhooks about losing students. My concern is for human rights and equity. 'Science for all' is not merely about opening the door of the science classroom to all students; it is about providing a science education that is valuable for everyone in the class, taking into account the needs and interests of individuals and groups. A science curriculum that is narrowly focussed and persistently committed to one way of thinking, one kind of knowledge, one position, privileges some students and denies effective learning to others. It drives some students out of science and draws others in, propagating and entrenching its particular view of science and students. It distorts education, it distorts science, and it distorts students. Education is not about dividing the kingdom, but building it.

Social judgement theories offer ways forward. First, students' anchor positions centre on values, principles or beliefs that they are prepared to defend. This means that the science classroom is not just about science; students' values need to be acknowledged, preferably early in a lesson sequence, and taken into account. Second, high levels of involvement shrink students' latitudes of no-commitment, and encourage systematic processing of information (as against heuristic processing). That involvement is quite self-centred. Hence participation, negotiation and power sharing help: as students express their interests and themselves and shape the classroom process, they build their involvement. Third, students' judgements are sensitive to social factors – other students' opinions, relationships in the group, wishes to seem consistent, concerns not to waste the current opportunity, and needs to give in order to receive. Groupwork and social interaction needs to be planned thoughtfully and done well. Fourth, students make judgements about incoming messages even as the messages come, and there are many messages. Students' judgements might result from cognitive analysis, attributional analysis, or heuristic thinking. In any case, intuition and perception are critical. This suggests that classroom processes should be fairly open, so that students' misgivings and untested attributions (and the teachers') can be raised and addressed. It will help also to have activities through which teachers and students all get to know each other in the work context, so that their intuitions are honed, and the classroom provides plenty of clues about what people are thinking and what they know and can do.

The social judgement process is probably more forgiving than my account above suggests, at least as it arises in classrooms. As noted earlier, the classroom is a captive space where teachers and students work together over time, building trust and getting to know each other. Over time, variety of teaching methods, acknowledgement of different positions, and use of different ways of knowing are all possible. In the give and take of the classroom, students are prepared to think in their non-preferred way today if they can use their preferred way tomorrow. What is more, they enjoy seeing how different students think. In my work with African school science students (Malcolm, 2004), students rated highly "learning from activities that can be done in different ways", "seeing how other students learn, and experimenting with different ways of learning" and "learning in groups where I can talk about problems and test my ideas". This suggests that Fatima's Rules are a fallback position: given content and processes that acknowledge and build on students' interests and needs, they will respond. Activities can easily be set up to highlight different ways of thinking and different positions, and make them part of lesson development.

Social judgement theory suggests a sequence, when working through the four ways of thinking in Herrmann's (1995) model: start with feelings thinking (C), in which students express their ideas and values, so that values, as well as knowledge, are on the agenda and ego-involvement is deepened; move to creative thinking (D), as students look for patterns and offer proposals, then to rational thinking (A), where proposals are analysed and tested, and finally to ordered thinking (B) where the knowledge is organised and shaped into rules and routines. This sequence is reminiscent of 'learning cycles' that have been proposed over the decades, by researchers such as Karplus, Novak and Bybee. What distinguishes it is that pushes more genuinely into the four ways of thinking, and gives them roughly equal time. For example, in the feelings-thinking segment, students work with feelings, think intuitively, look for non-verbal signals, and see relationships. In the creative-thinking segment, their thinking is adventurous, metaphorical, associative, impulsive, disruptive and free.

It follows that the definition of science in the curriculum is much broader than traditionally. It has to be to respond to more ways of thinking, more values positions, more diversity, than it chose to in the past. The resulting definitions are more like those in STS (Science, Technology, Society), Environmental Science and Multicultural Science, and more like science as it is practiced in a multitude of settings across the globe in the 21st Century.

Ultimately, the choice is a matter of purpose: whether the central purpose of science education is to select and train a group of students who think in a particular way, or to contribute to the science education of all students, in ways that make sense to them. Across the globe, science educators and scientists have successfully established science as a compulsory subject in general education, next only to Language and Mathematics in priority. This has been done in the name of 'science for all', implying that we intend science education to suit the full diversity of interests, abilities, thinking styles and cultures that exist. In practice, we have made some changes in teaching methods, but little in the content of the curriculum or the conception of science that underpins it. Almost as if by intention, we promote a position that is at odds with the positions of many students and they, as a consequence, place science in their latitudes of rejection or no-commitment. In the process, we divide students, un-necessarily, unhelpfully, into science and non-science groups, and encourage them to merely 'play the game' if they choose to persist with science at all. No matter which side of the science/non-science divide individuals fall, they are divided within, labelled and typed. And they have been denied rights to a science education that is valuable to them. We don't need to do that.

REFERENCES

Aikenhead, G. (1996). Science education: border crossing into the subculture of Science. *Studies in science education, 27*, 1-52.
Aikenhead, G. (2004). *Science education for everyday life: Evidence-based practice.* London, Ont.: Althouse Press.
Barber, M. (1996). National strategies for educational reform: Lessons from the British experience since 1988. In K. Leithwood, J. Chapman, D. Corson, P. Hallinger & A. Hart (Eds.), *International Handbook of Educational Leadership and Administration.* Dordrecht: Kluwer, 171-195.
Chaiken, S., Liberman, A., & Eagly, A. (1989). Heuristic and systematic information processing within and beyond the persuasion context. In J. Uleman & J. Bargh (Eds.), *Unintended thought.* New York: Guilford, 212-252.
Herrmann, N. (1995). *The creative brain.* Kingsport: Quebecor.
Horak, E., & du Toit, J. (2003). A study of the thinking styles and academic performance of civil engineering students. *CRREE Journal: For engineering students, 7*(1), 4-7.
Larson, J. O. (1995). Fatima's rules and other elements of an unintended chemistry curriculum. Paper presented to the American Educational Research Association, San Francisco, CA.
Malcolm, C. (2004). How do Grades 10-12 African students like to learn? In A. Buffler & R. C. Laugksch (Eds.), *Proceedings of the 12th Annual Conference of the Southern African Association for Research in Mathematics, Science and Technology Education.* Durban: SAARMSTE.
Malcolm, C., & Keogh, M. (2004). The science teacher as curriculum developer: Do you think it will rain today? In P. Naidoo, B. Gray & M. Savage (Eds.), *School science in Africa: teaching to learn, learning to teach.* Durban: AFCLIST/Juta, 105-128.
Mphahlele, E. (2002). *Es'kia.* Johannesburg: Kwela, in association with Stainbank and Associates.
Pooley, R. C., Anderson, G. K., Farmer, P., & Thornton, H. (1968). *England in Literature.* Glenview: Scott, Foresman and Company.
Posner, G. J., Strike, K. A., Hewson, P. W., & Gertzog, W. A. (1982). Accommodation of a scientific conception: Towards a theory of conceptual change. *Science Education, 66*(2), 211-227.
Sherif, M., Sherif, K., & Nebergall, R. (1965). *Attitude and attitude change: The social judgment-involvement approach.* Philadelphia: Saunders.
Sjøberg, S. (2002). Science and technology education: Current challenges and possible solutions. In E. Jenkins (Ed.), *Innovations in Science and Technology Education* (vol. VIII). Paris: UNESCO.
Snow, C. P. (1959). *Two cultures and the scientific revolution.* Cambridge: Cambridge University Press.

Steel, D. (2002). Danielle Steel. In B. Conrad & M. Schulz (Eds.), *Snoopy's Guide to the Writing Life.* Cincinnati: Writer's Digest Books.

Tarnas, R. (1993). *The passion of the Western mind.* New York: Ballantine.

AFFILIATIONS

Cliff Malcolm
Department of Science Education,
University of KwaZulu-Natal

ALAN J. BISHOP

VALUES IN MATHEMATICS
AND SCIENCE EDUCATION

An Empirical Investigation

This chapter describes the genesis, background and some data from a recently completed research project on values in mathematics and science education. Values are theorised as being the deep affective qualities of the subjects which are revealed through the educational process. This project involved two mathematics and two science educators and the research was carried out with teachers and their students in both primary and secondary schools. Here we show that although there are some strong similarities between the values ascribed in the research literature to the two subject areas, there are some important differences perceived by the educators in those fields. From the other data we can see that differences are perceived by the teachers, although not always the same differences. Nor do the students agree with their teachers about the values priorities.

INTRODUCTION AND BACKGROUND

In the modern knowledge economy, societies are demanding greater mathematical and scientific literacy and expertise from their citizens than ever before. At the heart of such demands is the need for greater engagement by students with school mathematics and science. As the OECD/PISA definition of numeracy puts it:

> Mathematical literacy is an individual's capacity to identify and understand the role that mathematics plays in the world, to make well-founded judgements and to use and engage with mathematics in ways that meet the needs of that individual's life as a constructive, concerned and reflective citizen. (OECD, 2003: 10)

Values are an inherent part of the educational process at all levels, from the systemic, institutional macro-level, through the meso-level of curriculum development and management, to the micro-level of classroom interactions (Le Métais, 1997) where they play a major role in establishing a sense of personal and social identity for the student.

The notion of 'values' is not new in anthropology (e.g., Kluckholm, 1962), in psychology (e.g., Kohlberg, 1981; Krathwohl, Bloom, & Masia, 1964; Rokeach, 1973), or in general education (e.g., Halstead, 1996; Nixon, 1995; Raths, Harmin, & Simon, 1987). However the notion of studying values in mathematics education

U. Gellert, E. Jablonka (eds.), *Mathematisation – Demathematisation: Social, Philosophical and Educational Ramifications, 123–139.* © 2007 Sense Publishers. All rights reserved.

is a relatively recent phenomenon (Bishop, 1999). According to Chin, Leu and Lin (2001), the values portrayed by teachers in mathematics classrooms are linked to their pedagogical identities. Seah and Bishop (2001) describe the values held by teachers as representing their 'cognitisation' of affective variables such as beliefs and attitudes, and the subsequent internalisation of these values into their respective affective-cognitive personal system.

Even in science education the study of values in classrooms is not a major focus of research. Nevertheless, in mathematics and science education values are crucial components of classrooms' affective environments, and thus have a crucial influence on the ways students choose to engage (or not engage) with mathematics and science. Clearly the extent and direction of this influence will depend on the teachers' awareness of, respectively, values ascribed to the particular discipline, the values carried by their selection from the available pedagogical repertoire, and their consciousness or otherwise of imposing their own personal values (Pritchard & Buckland, 1986).

Data from a previous research project, the Values and Mathematics Project (VAMP), has shown that teachers of mathematics are rarely aware of the values associated with teaching mathematics (FitzSimons, Seah, Bishop, & Clarkson, 2000)[52].

Furthermore, any values 'teaching' which does occur during mathematics classes happens implicitly rather than explicitly (Bishop, 2002). This chapter will report on ideas developed from a more recent research project concerned with values in both mathematics and science education. There were three basic questions on which the research focused:

- (1) What values are (a) implicit and (b) explicit in the intended mathematics and science curriculum and assessment documents, as well as in textual and other resources utilised in the classrooms under study?
- (2) What values do teachers (a) espouse socio-historically, epistemologically, and pedagogically and (b) actually portray in their classrooms? How do each of these sets of values relate to those in (1)?
- (3) What values do students in these classrooms hold, and how do these relate to the values in each of (1) and (2)? How, if at all, are they influenced by pedagogical interventions within the timeframe of the school year?

This project therefore differed from the earlier one in two significant ways. Firstly it involved a comparison between values in mathematics and science. This was done because of both the similarities between the two subjects and their differences. The second difference with VAMP was that this one would also involve collecting data from students, and would attempt to explore their values and how these are related to any the teachers might hold. That aspect was to prove very challenging.

THEORETICAL FRAMEWORK

Comparing values teaching in different subject areas is a relatively novel research approach and some parallel research on teachers of mathematics and history by

Bills and Husbands (2004), which builds on the ideas of Gudmunsdottir (1991) from English and history teachers, also shows what can be learnt from this approach. An important perspective on values, of relevance to this present study, is offered by Billett's (1998) analysis of the social genesis of knowledge. His analysis points to the different sources of influence on teachers' values. Billett categorises knowledge at five levels, and below is an indication of how different knowledge at these levels can impinge on and influence teachers' values.

- (1) *Socio-historic knowledge* factors affect the values underpinning decisions made by both management and teachers. For example, we can find important values ideas in the writings of Popkewitz (1998) concerning the historic, ideological purposes of schooling.
- (2) *Socio-cultural practice* is described by Billett as historically derived knowledge transformed by cultural needs; goals, techniques, and norms to guide practice; and expectations of transformed socio-historic knowledge. These are manifested by curricular decisions influenced by such factors as state or national curricular frameworks such as the *Victorian Curriculum and Standards Framework for Mathematics* (Board of Studies, 1999).
- (3) *The community of practice in the classroom* is identified by Billett as particular sociocultural practices shaped by a complex of circumstantial social factors (activity systems), and the norms and values which embody them.
- (4) *Microgenetic development* is interpreted by Billett as individuals' (teachers' and students') moment-by-moment construction of socially derived knowledge, derived through routine and non-routine problem solving.
- (5) Billett's last category is *ontogenetic development*, in which he included individuals' personal life histories. Thus the values teaching in the mathematics classroom would be likely to be influenced by: (a) the teacher's prior experiences of learning mathematics, researching mathematics education, classroom teaching, and using mathematics in other life/workplace experiences; and (b) the students' prior experiences of using and learning mathematics in formal, informal, and non-formal settings.

The first part of this chapter is particularly concerned with ideas and influences at Billett's levels 1, 2 and 3, and the latter part with how these play out in the actual classroom practices in levels 4 and 5.

It was decided that for this study, in order to have some basis for the mathematics and science comparisons it would be necessary to develop a theoretical framework for the values studied. We used the six values cluster model developed by one of the authors (Bishop, 1988), based on his analysis of the writings concerning the activities of mathematicians throughout Western history and culture. It is important to stress that the emphasis in that analysis was not primarily on which values might be, are, or should be, emphasised in mathematics education, but rather on the development of mathematics as a subject throughout Western history.

In this model, six value clusters are structured as three complementary pairs, related to the three dimensions of ideological values, sentimental values, and sociological values. These three dimensions are based on the original work of

White (1959), a renowned culturologist, who proposed four components to explain cultural growth. White nominated these as technological, ideological, sentimental (or attitudinal), and sociological, with the first being the driver of the others. Bishop (1988) argued that mathematics could be considered as a symbolic technology, representing White's technological component of culture, with the other three being considered as the values dimensions driven by, and also in their turn driving, that technology.

The six value clusters that Bishop (1988) originally identified are described as follows:

> The particular societal developments which have given rise to Mathematics have also ensured that it is a product of various values: values which have been recognised to be of significance in those societies. Mathematics, as a cultural phenomenon, only makes sense if those values are also made explicit. I have described them as complementary pairs, where *rationalism* and *objectism* are the twin ideologies of Mathematics, those of *control* and *progress* are the attitudinal values which drive Mathematical development, and, sociologically, the values of *openness* and *mystery* are those related to potential ownership of, or distance from Mathematical knowledge and the relationship between the people who generate that knowledge and others. (Bishop, 1988: 82)

VALUES IN MATHEMATICS AND SCIENCE

Regarding their similarities, both mathematics and science are taken as ways of understanding that are embedded in rational logic – focusing on universal knowledge statements. Both are seen by society in general as essential components of schooling, rivalled only by literacy. Hence, teachers of each face substantial political and social pressures from outside the school (e.g., system-wide assessments of student performance, purposes for teaching seen as being directly related to technological development, etc.). In their teaching, both involve following routines, although not exclusively. Both involve modelling, albeit with different emphases. Similarly each is incorporated into the other's applications but in an asymmetrical relationship.

On the other hand, science curricula/texts commonly contain a section on "The Nature of Science" while mathematics rarely contains the equivalent. While the values embedded in mathematics teaching are almost always implicit, in science teaching some are quite explicit. For example, curriculum movements such as *Science-Technology-Society* make some values explicit and central to the intended learning outcomes; laboratory work seeks to make explicit such values as 'open mindedness,' 'objectivity,' etc.; and content described as *The Nature of Science*, for example, also makes some values explicit (see also UNESCO, 1991).

Among the general public, although the concept of 'a science industry' or 'scientific industries' is widely recognised, this is not the case for mathematics. In the popular media (e.g., magazines, newspapers, books, radio, television), science receives much more attention than mathematics, despite there having been a few recent movies featuring mathematical prodigies. Even when it is present, mathematics is generally subsumed under science. In the case of the popular pursuit of gambling, where mathematical thinking might be considered to play an important role, this is generally not the case as 'luck' seems to be considered a critical factor for many people.

Yet mathematics plays a much more prominent role as a gatekeeper in society than does science. For example, it is often used as a selection device for entry to higher education or employment, even when the skills being tested are unrelated to the ultimate purpose. In broad terms (e.g. modelling or simulations which reduce costs and/or danger), mathematics is considered to be publicly important; at the very same time as it is considered to be personally irrelevant (Niss, 1994), apart from the obvious examples of cooking, shopping and home maintenance. Politically, mathematics has been ascribed a *formatting role* in society (Skovsmose, 1994).

DIFFERENCES IN VALUES BETWEEN MATHEMATICS AND SCIENCE

As has been said above, this project used as its basic conceptual framework the six values component model developed by the author (Bishop, 1988). In this model six sets of value clusters are structured as three complementary pairs.

Table 1. Values of Western mathematical knowledge (Bishop, 1988)

1. Epistemology of the knowledge (Ideological values)	
1a Rationalism	*1b Objectivism*
Reason; Explanations; Hypothetical reasoning; Abstractions; Logical thinking; Theories	Atomism; Objectivising; Materialism; Concretising; Determinism; Symbolising; Analogical thinking

2. How individuals relate to the knowledge (Sentimental values)	
2a Control	*2b Progress*
Prediction; Mastery over environment; Knowing; Rules; Security; Power	Growth; Questioning; Alternativism; Cumulative development of knowledge; Generalisation

3. Knowledge and society (Sociological values)	
3a Openness	*3b Mystery*
Facts; Universality; Articulation; Individual liberty; Demonstration; Sharing; Verification	Abstractness; Wonder; Unclear origins; Mystique; Dehumanised knowledge

The project involved two mathematics educators and two science educators, and in the first part of the project there was considerable discussion and analysis of this initial framework, particularly in relation to whether the same structure could hold for science (see Corrigan, Gunstone, Bishop, & Clarke, 2004). As a result of this analysis, a comparison of values between the mathematics and science educators was achieved, as shown in *Table 2*.

Table 2. Comparison between values associated with mathematics and science

Mathematics	Science
Rationalism Reason; Explanations; Hypothetical reasoning; Abstractions; Logical thinking; Theories	*Rationalism* Reason; Explanations; Hypothetical reasoning; Abstractions; Logical thinking; Theories
Empiricism Atomism; Objectivising; Materialism; Concretising; Determinism; Symbolising; Analogical thinking	*Empiricism* Atomism; Objective; Materialisation; Symbolising; Analogical thinking; Precise; Measurable; Accuracy; Coherence; Fruitfulness; Parsimony; Identifying problems
Control Prediction; Mastery over environment; Knowing; Rules; Security; Power	*Control* Prediction; Mastery over problems; Knowing; Rules; Paradigms; Circumstance of activity
Progress Growth; Questioning; Cumulative development of knowledge; Generalisation; Alternativism	*Progress* Growth; Cumulative development of knowledge; Generalisation; Deepened understanding; Plausible alternatives
Openness Facts; Universality; Articulation; Individual liberty; Demonstration; Sharing; Verification	*Openness* Articulation; Sharing; Credibility; Individual liberty; Human construction
Mystery Abstractness; Wonder; Unclear origins; Mystique; Dehumanised knowledge; Intuition	*Mystery* Intuition; Guesses; Daydreams; Curiosity; Fascination

As can be seen there is a considerable amount of agreement, but there are some important differences. As far as the Ideological dimension is concerned there are both similarities and differences. In the cluster of Rationalism there is much agreement, as both subjects require the use of all the logic skills available and thus emphasise the range of values associated with those skills. With the value cluster of Objectism, which became recast as 'Empiricism' in order to accommodate the scientists' approach, there is also some agreement, but the highly empirical nature of science means that it has many more value aspects there than does mathematics. The experimental and observational activities of science bring other values into play than we can find in doing mathematics.

For the Sentimental dimension, with the complementary pairing of Control and Progress, there was once again some agreement between the mathematics and science educators about the Control value cluster, with its emphasis on prediction, mastery, and procedural rules. However the circumstances of the activity and different paradigms are significant in science but have little meaning in mathematics. Likewise with Progress, the idea of the cumulative development of knowledge is clearly similar, but the role of science in continuing to deepen understanding of a phenomenon again has no parallel in mathematics development.

Some other differences appear with the Sociological dimension, that is the way individuals relate to the knowledge of the subject. In relation to the Openness value cluster, the emphasis of science on credibility and human construction are significant, compared with the idea of 'facts' in mathematics and the value of verification, sometimes via proof. With Mystery, which itself is a rather mysterious category, the dehumanised nature of mathematics with its abstractness and unclear notions of the origins of ideas contrasts strongly with the intuition, daydreaming, and empirically-based guesses of the scientists.

When considering these contrasts it is important to remember that this framework involves pairs of clustered values along the three dimensions. So the two clusters should not be considered as dichotomous, but rather as complements of each other. For example, Openness is the complement of Mystery, and therefore both clusters are present to some extent in that value dimension. Furthermore, what the model suggests is not that science and mathematics are markedly different but that there are strong similarities in their values, as befits their common heritage. There are however some interesting and, in terms of education, revealing different values represented also.

TEACHERS' VALUES AND PRACTICES

We now turn to some of the data collected from the primary and secondary teachers by means of specially constructed questionnaires. They were based on the three complementary pairs, Rationalism and Empiricism, Control and Progress, Openness and Mystery, discussed above. The same structure was used for the mathematics and the science questionnaires and for the primary and secondary teachers, although there were some minor adjustments in the descriptions of teaching situations. 13 primary teachers of years 5/6 and 17 secondary teachers of years 7/8 volunteered to answer these questionnaires. Primary teachers in the state system in Australia teach both subjects to their classes, and we also chose secondary teachers who taught both subjects to the same classes.

Questions 1 and 2 of the questionnaires (see the Appendix) ask for the extent to which particular activities are emphasised in practice in the teacher's mathematics (and science) classes. The items in these questions are designed to explore, in sequence, aspects of Rationalism and Empiricism, Control and Progress, Openness and Mystery. So, in the Appendix it can be seen that the first three statements in Question 1 all relate to the value of Rationalism, and so on through the 18 items in Question 1.

Question 2 uses the same structure (a group of 3 items relating to each of the 6 values in the three pairs) to ask about the frequency of use of specific classroom activities. In the Appendix can be seen the various statements, but not in the actual format used in the questionnaires.

For all the statements in Questions 1 and 2, we scored the responses as 4 (for "Always"), 3 ("Often"), 2 ("Sometimes"), 1 ("Rarely"), and also calculated means. We recognise that in doing this we have taken an ordinal scale and treated it as if it was a ratio scale.

To facilitate comprehension of the results, we have combined the data for Questions 1 and 2, and in the data reported below, for example, a teacher's view of the frequency of emphasis on Rationalism in his/her class' activities is represented by the mean score for the six items relating to that value cluster in the two questions.

Questions 3 and 4 are the questions which concern the teachers' preferences for the six value clusters described above. The structure of these questions is that each question contains 6 statements to be ranked by the teachers. Each statement relates to one of the values clusters, for example, the statement "It develops creativity, basing alternative and new ideas on established ones" relates to the value of Progress. The other statements follow closely the other value descriptors although their order is different in the two questions. Note also that although the teachers knew we were studying values, they were not made aware of the value structure underlying the questions and the various statements.

Tables 3-6 show the results from the two groups of teachers in terms of their rankings of the six values clusters. In brackets are the means of (a) the frequencies in Questions 1 and 2, and (b) the rank orders in Questions 3 and 4.

Table 3. Teachers' preferred values and their preferred teaching practices: rank orders: primary mathematics

	Rationa- lism	Empiri- cism	Control	Progress	Open- ness	Mystery
Qus. 1/2	4 (2.64)	2 (2.80)	1 (2.95)	5 (2.44)	3 (2.65)	6 (2.25)
Qu. 3	2 (2.30)	1 (1.46)	6 (5.23)	4 (3.15)	3 (3.53)	5 (3.61)
Qu. 4	3 (3.66)	1 (1.33)	5 (3.75)	2 (3.00)	3 (3.66)	6 (3.83)

We can see from Table 3 that there is a close similarity between the primary teachers' views on questions 3 and 4, and some close correlation between them and questions 1/2 particularly regarding Empiricism, Openness and Mystery. However, the ranks for Control stand out as being markedly different.

Table 4. Teachers' preferred values and their preferred teaching practices: rank orders: primary science

	Rationa-lism	Empiri-cism	Control	Progress	Open-ness	Mystery
Qus. 1/2	2 (3.05)	3 (2.90)	1 (3.07)	4 (2.57)	5 (2.47)	6 (1.91)
Qu. 3	2 (2.75)	1 (1.41)	6 (4.91)	4 (3.41)	5 (3.66)	3 (3.00)
Qu. 4	4 (3.41)	1 (1.41)	6 (4.75)	3 (3.33)	5 (3.83)	2 (2.58)

For Science the primary teachers again express similar views for Questions 3 and 4, and once again the ranks for Control are markedly different from that in Questions 1/2. Mystery is also ranked differently in practice from the teachers' preferred views.

Table 5. Teachers' preferred values and their preferred teaching practices: rank orders: secondary mathematics

	Rationa-lism	Empiri-cism	Control	Progress	Open-ness	Mystery
Qus. 1/2	2 (2.15)	3 (2.05)	1 (2.75)	5 (1.93)	4 (1.99)	6 (1.79)
Qu. 3	1 (1.94)	2 (2.05)	6 (4.52)	4 (3.88)	3 (3.35)	5 (4.29)
Qu. 4	1 (1.70)	2 (1.82)	3 (3.44)	4 (4.00)	4 (4.00)	6 (4.47)

The secondary teachers rank Rationalism highest for mathematics in terms of their preferred values (Questions 3 and 4) but, like their Primary colleagues, they place Control in the highest rank in practice.

Table 6. Teachers' preferred values and their preferred teaching practices: rank orders: secondary science

	Rationa-lism	Empiri-cism	Control	Progress	Open-ness	Mystery
Qus. 1/2	1 (2.86)	3 (2.61)	2 (2.84)	5 (2.30)	4 (2.33)	6 (2.03)
Qu. 3	4 (3.18)	1 (1.25)	6 (5.87)	4 (3.18)	3 (3.06)	2 (2.81)
Qu. 4	3 (3.12)	1 (1.25)	6 (4.12)	2 (3.00)	5 (4.06)	4 (3.33)

For the secondary teachers and science, Questions 3 and 4 show us that the teachers' main value preference is for Empiricism, but in practice they favour Rationalism with Control coming a close second. Once again we can see differences with respect to Control, but this time also with Mystery.

The comparisons between the values in mathematics and science for the two groups of teachers show interesting differences, reflecting their concerns with the curriculum and teaching at their respective levels. For the primary teachers, concerning Ideology, they prefer Empiricism over Rationalism for both science and mathematics, though both are important, rankings which are also reflected in the findings for their preferred practices. At the primary level of course much mathematical work is empirical in nature. For the Sentimental dimension, Control is much less favoured than Progress also for both, but the practices are very different. Another main difference between the subjects appears in the Sociological dimension where Openness and Mystery reverse their positions with the two subjects, the first being more favoured than the second in mathematics and the reverse in science. This difference does not translate to the practices however, with the science practices being ranked much more like the mathematics practices.

For the secondary teachers, concerning the Ideological dimension, they favour Rationalism for mathematics and Empiricism for science, disagreeing with the primary teachers. For the Sentimental dimension, the secondary teachers largely agree with their primary colleagues and for the Sociological dimension, they again agree with their primary colleagues favouring Openness for mathematics compared with Mystery, and reversing these for science. Indeed Mystery for science is ranked 2 and 4 by the secondary teachers and ranked 2 and 3 by the primary teachers, showing how significant they consider that aspect to be.

COMPARISON OF PRIMARY TEACHERS' AND STUDENTS' VALUE PREFERENCES

In this part of the chapter we report on some of the differences and similarities between the primary teachers' and their students' data. Regrettably it proved impossible to obtain similar student data from the secondary teachers. Whether they were reluctant to allow their students to answer the questionnaire we do not know. Various reasons were given, chiefly in relation to the time necessary to administer the questionnaire (20 minutes maximum!).

In this first attempt to learn about the students' values it was decided to give them a version of the questionnaire which we used with the teachers, with the language suitably modified for the young learners (see the Appendix). So, regarding the Question 3 for mathematics we can see in *Table 7* several differences between the primary teachers' and their students' rank orders. Indeed there is almost a kind of reversal with the teachers' 1 and 2 being the students' 5 and 6, and almost vice-versa. Of course there are differences in the wording of the two sets of statements which could easily have had confounding effects.

Table 7. Rank orders for the values in Question 3 between primary teachers and primary students (mathematics)

	Rationalism	Empiricism	Control	Progress	Openness	Mystery
Teachers	2 (2.30)	1 (1.46)	6 (5.23)	4 (3.15)	3 (3.53)	5 (3.61)
Students	6 (4.04)	5 (3.97)	3 (3.37)	1 (2.72)	4 (3.85)	2 (3.11)

For the students' value orders in Question 4 (mathematics), *Table 8* shows that Control stands out in the first place, with Rationalism second. This is in marked contrast to the responses to Question 3 above. The reasons are not clear.

Table 8. Rank orders for the values in Question 4 between primary teachers and primary students (mathematics)

	Rationalism	Empiricism	Control	Progress	Openness	Mystery
Teachers	3 (3.66)	1 (1.33)	5 (3.75)	2 (3.00)	3 (3.66)	6 (3.83)
Students	2 (3.20)	4 (3.65)	1 (2.54)	3 (3.38)	6 (4.23)	5 (3.97)

For the values in Question 3 in Science (*Table 9*) the clear first choice for the students was Progress ("We get to invent new ideas") as it was for mathematics in *Table 7*. There seems to have been some interesting teaching going on in these classes!

Table 9. Rank orders for the values in Question 3 between primary teachers and primary students (science)

	Rationalism	Empiricism	Control	Progress	Openness	Mystery
Teachers	2 (2.75)	1 (1.41)	6 (4.91)	4 (3.41)	5 (3.66)	3 (3.00)
Students	4 (3.64)	2 (3.07)	5 (3.65)	1 (2.88)	6 (4.39)	3 (3.40)

For the students' values in Question 4 in Science (*Table 10*) once again, Progress comes out on top, with Control in the second place, different from their teachers. Openness is again an unfavoured value for the students.

Table 10. Rank orders for the values in Question 4 between primary teachers and primary students (science)

	Rationalism	Empiricism	Control	Progress	Openness	Mystery
Teachers	4 (3.41)	1 (1.41)	6 (4.75)	3 (3.33)	5 (3.83)	2 (2.58)
Students	3 (3.45)	5 (4.05)	2 (2.98)	1 (2.96)	6 (4.22)	4 (3.48)

Empiricism, which was always the strongest number 1 preferred value for the primary teachers was never number 1 for the students, although it did get to number 2 in *Table 8* above. In the other tables Empiricism was low down in the list.

Another interesting feature is the difference between teachers and students with respect to the Control value. We can see in *Tables 7-10* that for the teachers it is a low ranking while for their students it ranks much higher. So we may like to think that this shows that the teachers' values don't influence those of their students. But recall the data in the previous section. We saw in *Tables 3-6* that although they ranked Control as a low priority in questions 3 and 4 (their preferred values), when we look at the data from Questions 1 and 2, reflecting their actual teaching, the values *in practice* certainly do put Control on top (with only one exception)!

It appears, in this case, that the students are learning this value from the teachers' actions and behaviours in class, and not from what the teachers' preferences are.

CONCLUSIONS AND IMPLICATIONS

The comparison of the values between the science and mathematics educators in the project has revealed perceptions of some important differences between the two subjects. It has also helped to clarify the values structure underlying the current project. In particular, regarding the Ideological dimension, there was evidence that mathematics educators favour the cluster of Rationalism while science educators emphasise Empiricism.

With the Sentimental dimension, while both subjects favour Control, the values of Progress differ, with science seeking to deepen understanding of relationships rather than constructing new knowledge as in mathematics. Concerning the Sociological dimension, there are important differences in both the Openness and Mystery clusters with science seeming to relate more to the humanising aspects of knowledge compared with mathematics.

The comparisons between the values in mathematics and science for the teachers also show interesting differences, reflecting their concerns with the curriculum and teaching at their respective levels. At the primary level the teachers favour Empiricism over Rationalism for both science and mathematics, though both are important, and this contrasts with the findings above. At the primary level of course much mathematical work is empirical in nature. For the Sentimental dimension, Control is much less favoured than Progress also for both. The main difference between the subjects appears in the Sociological dimension where Openness and Mystery reverse their positions with the two subjects, the first being more favoured than the second in mathematics and the reverse in science. This difference shown by the primary teachers reflects the educational implications of the educators' views above.

For the secondary teachers, the Ideological dimension reflects the educators' views, with mathematics favouring Rationalism and science favouring Empiricism, disagreeing with the primary teachers. For the Sentimental dimension, the secondary teachers largely agree with their primary colleagues and for the Sociological dimension, they again agree with their primary colleagues favouring Openness for mathematics compared with Mystery, and reversing these for science. Indeed Mystery for science is ranked 2 and 4 by the secondary teachers and ranked 2 and 3 by the primary teachers, showing how significant they consider that aspect to be.

For the primary students there is evidence that their priorities in the value statements are largely unrelated to the teachers' stated value preferences. However there is limited evidence, in the case of the value Control, that the students' higher ranking of this value coincides with the teachers' actual practices and not their preferences.

In general, the conceptualisation put forward for this project has begun to show interesting and interpretable results. Discussions with the teachers have revealed an interest in the issues of values teaching in all subjects, but also a lack of vocabulary, and conceptual tools to enable them to develop explicitly the values underlying mathematics education. One of the goals of this project was by contrasting mathematics and science, to help teachers develop those conceptual tools further. As we have seen, and as has been shown above, the contrasts between these two closely related forms of knowledge are provocative, and already reveal worthwhile challenges particularly for mathematics teaching to pursue.

For example, the difference between the emphasis on Empiricism at primary level and on Rationalism at secondary level implies some important challenges for explicit values development in the teaching of mathematics at those two levels. How should that values development be smoothed across the primary/secondary divide?

The differences in the views on Progress are also revealing, with the development of understanding in science contrasting with the construction of new knowledge in mathematics. How can we reconstruct our views of the mathematics curriculum so that progress through that curriculum is not just a matter of acquiring new knowledge but of ensuring that it also deepens learners' understanding of what has been taught before?

Finally could the dehumanised, highly abstract and mystique-laden value of the Mystery of mathematics which appears to be such an obstacle to mathematics learners be made more explicit so that it could be challenged by the more humanised and personal intuitive nature of that value which science appears to enjoy?

However, before jumping to too many conclusions, we must remember that the data are from questionnaires and consist of teachers' reported views of their preferences and their practices. We do not know the extent to which their rankings of these practice statements reflect their actual practices. However, the data for science at the secondary level, where teachers emphasise other values than for mathematics, indicates the usefulness of comparing subjects and their values emphases.

Finally one can see that, if the data reported here are valid, the differences show that teachers' values in the classroom are shaped to some extent by the values embedded in each subject, as perceived by them. This implies that changing teachers' perceptions and understandings of the subject being taught may well change the values they can emphasise in class. Further if teachers wish to emphasise values other than those they currently emphasise, it is possible to learn strategies from their teaching of other subjects. However when it comes to influencing students' values, teachers probably should look to their classroom practices as the main influence. There is clearly a need to explore more the relationship between teachers' practices and their students' values.

ACKNOWLEDGEMENTS

Thanks are due to my colleagues Debbie Corrigan, Barbara Clarke, and Dick Gunstone for their contributions to this project and to this chapter.

NOTES

[52] Various relevant papers from that study, and from other authors, are available from this website: http://www.education.monash.edu.au/centres/scienceMTE/vamppublications.html

REFERENCES

Billett, S. (1998). Transfer and social practice. *Australian and New Zealand Journal of Vocational Education Research, 6*(1), 1-25.

Bills, L., & Husbands, C. (2004). Analysing of embedded values in history and mathematics classrooms. Paper presented at the British Educational Research Association Conference in Manchester, September 2004.

Bishop, A. J. (1988). *Mathematical enculturation: A cultural perspective on mathematics education.* Dordrecht: Kluwer.

Bishop A. J. (1999). Mathematics teaching and values education – an intersection in need of research. *Zentralblatt für Didaktik der Mathematik, 31*(1), 1-4.

Bishop, A. J. (2002). Research policy and practice: the case of values. Paper presented to the Third conference of the Mathematics Education and Society Group, Helsingør, Denmark, April 2002.

Board of Studies (1999). *Mathematics: Curriculum and standards framework I.* Carlton: Board of Studies.

Chin, C., Leu, Y.-C., & Lin, F.-L. (2001). Pedagogical values, mathematics teaching, and teacher education: Case studies of two experienced teachers. In F.-L. Lin & T. J. Cooney (Eds.), *Making sense of mathematics teacher education.* Dordrecht: Kluwer, 247-269.

Corrigan, D. J., Gunstone, R. F., Bishop, A. J., & Clarke, B. (2004). Values in science and mathematics education: Mapping the relationships between pedagogical practices and student outcomes. Paper presented at Summer School of the European Science Educational Research Association, Mühlheim, Germany, August 2004.

FitzSimons, G. E., Seah, W. T., Bishop, A. J., & Clarkson, P. C. (2000). What might be learned from researching values in mathematics education? In T. Nakahara & M. Koyama (Eds.), *Proceedings of the 24th conference of the International Group for the Psychology of Mathematics Education* (vol. 1). Hiroshima: Hiroshima University, 153.

Gudmunsdottir, S. (1991). Values in pedagogical content knowledge. *Journal of Teacher Education, 41*(3), 44-52.

Halstead, M. (1996). Values and values education in schools. In J. M. Halstead & M. J. Taylor (Eds.), *Values in education and education in values.* London: Falmer, 3-14.

Kluckholm, C. (1962). *Culture and behavior.* New York: Macmillan.

Kohlberg, L. (1981). *The philosophy of moral development: Moral stages and the idea of justice.* San Francisco: Harper & Row.

Krathwohl, D. R., Bloom, B. S., & Masia, B. B. (1964). *Taxonomy of educational objectives: The classification of educational goals (Handbook II: Affective domain).* New York: McKay.

Le Métais, J. (1997). *Values and aims underlying curriculum and assessment.* (International Review of Curriculum and Assessment Frameworks Paper 1). London: School Curriculum and Assessment Authority.

Niss, M. (1994). Mathematics in society. In R. Biehler, R. W. Scholz, R. Sträßer, & B. Winkelmann (Eds.), *Didactics of mathematics as a scientific discipline.* Dordrecht: Kluwer, 367-378.

Nixon, J. (1995). Teaching as a profession of values. In J. Smyth (Ed.), *Critical discourses on teacher development.* London: Cassell, 215-224.

OECD (2003). *Assessment framework – mathematics, reading, science and problem solving: Knowledge and skills.* Paris: OECD.

Popkewitz, T. S. (1998). *Struggling for the soul: The politics of schooling and the construction of the teacher.* New York: Teachers College Press.

Pritchard, A. J., & Buckland, D. J. (1986). *Leisure, values and biology teaching* (Science and Technology Education, Document Series No. 22) Paris: UNESCO.

Raths, L. E., Harmin, M., & Simon, S. B. (1987). Selections from 'values and teaching'. In J. P. F. Carbone (Ed.), *Value theory and education.* Malabar: Robert E. Krieger, 198-214.

Rokeach, M. (1973). *The nature of human values.* New York: The Free Press.

Seah, W. T., & Bishop, A. (2001). Teaching more than numeracy: The socialization experience of a migrant teacher. In J. Bobis, B. Perry & M. Mitchelmore (Eds.), *Numeracy and beyond.* Turramurra: MERGA, 442-450.

Skovsmose, O. (1994). *Towards a philosophy of critical mathematics education.* Dordrecht: Kluwer.

UNESCO (1991). *Values and ethics and the science and technology curriculum.* Bangkok: Asia and the Pacific Programme of Educational Innovation for Development, UNESCO-729.

White, L. A. (1959). *The evolution of culture.* New York: McGraw-Hill.

APPENDIX

Questions from the teachers' questionnaire (secondary mathematics version)

Qu.1. When you are teaching mathematics to Years 7 and 8, how often do you emphasise the following?
– How often do you emphasise the role of proving in mathematics?
– How often do you have structured debates in class?
– How often do you encourage your students to argue seriously with each other in your classes?
– How often do you use diagrams to illustrate mathematical relationships?
– How often do you encourage your students to invent their own symbols and terminology *before* showing them the 'official' ones?
– How often do you use concrete materials (e.g. physical models) to demonstrate mathematical relationships?
– How often do you emphasise the checking of right answers, *and* the reasons for other answers not being 'right'?
– How often do you encourage the analysis and understanding of *why* routine calculations and algorithms 'work'?
– How often do you show examples of how the mathematical ideas you are teaching are used in the real world?
– How often do you encourage alternative, and non-routine, solution strategies together with their reasons?
– How often do you encourage students to extend and generalise ideas from particular examples?
– How often do you give the students stories and examples of recent mathematical developments?

137

- How often do you encourage your students to defend and justify their answers and methods publicly to the class?
- How often do your students create posters to display their ideas to the others?
- How often do you demonstrate how mathematical ideas can be shown to be true?
- How often do you stimulate your students' mathematical imagination with pictures, artworks, etc.?
- How often do you use mathematical puzzles in class?
- How often do you tell students stories about mathematical discoveries?

Qu. 2. How frequently do you use any of these activities in your mathematics teaching at this level?
- Small group discussions
- Whole class discussions
- Investigations
- Modelling activities
- Using manipulatives
- Role playing real-life situations
- Practising algorithms
- Memorising facts
- Problem solving
- Generating conjectures and hypotheses
- Having students generate questions
- Historical and cultural projects
- Students explaining at the board
- Students making posters and displays
- Proving generalisations
- Displaying famous 'mathematical' artwork
- Mathematical puzzles
- Using mathematical paradoxes

For the next two items please rank the six statements accordingly in the accompanying boxes, where '1' indicates your first choice, '2' your second choice, '3' your third choice, etc. Note that the same ranking value can be given to more than one statement. Please rank each statement.

Qu. 3. "For me, Mathematics is valued in the school curriculum because...."

It develops creativity, basing alternative and new ideas on established ones ☐

It develops rational thinking and logical argument ☐

It develops articulation, explanation and criticism of ideas ☐

It provides an understanding of the world around us ☐

It is a secure subject, dealing with routine procedures and established rules ☐

It emphasises the wonder, fascination and mystique of surprising ideas ☐

Qu. 4. "For me, Mathematics is valuable knowledge because..."

It emphasises argument, reasoning and logical analysis ☐

It deals with situations and ideas that come from the real world ☐

It emphasises the control of situations through its applications ☐

New knowledge is created from already established structures ☐

Its ideas and methods are testable and verifiable ☐

It is full of fascinating ideas which seem to exist independently of human actions ☐

Questions 3 and 4 from the primary student questionnaire.

I like maths lessons because: (rank order)

We get to argue with each other ☐

We do lots of practical work ☐

We try to solve real problems ☐

We get to invent new ideas ☐

We get to show the others how we do things ☐

We do lots of puzzles ☐

Maths is important for my future because: (rank order)

It helps me to think ☐

Its about solving real problems ☐

It teaches me lots of useful skills ☐

It makes me be creative ☐

I learn to tell others about my ideas ☐

It shows me that all kinds of problems are interesting ☐

AFFILITATIONS

Alan J. Bishop
Faculty of Education,
Monash University, Melbourne

AFZAL AHMED

SOME EXPLORATIONS INTO INHIBITORS AND FACILITATORS IN LEARNING MATHEMATICS

I will base my explorations around the themes of my many interesting and absorbing conversations with Christine Keitel over more than twenty years. It is therefore appropriate to begin by quoting her.

> while common sense is bound to the context, and meant for the immediate use, mathematical abstraction as the constituent of mathematics [...] consistently tends to become context-free and universal. [...] By the separation of mathematics from natural sciences and human endeavours, by dismissing application and legitimation problems as well as social accountability in the course of specialisation, by concentrating on the formal refinement of prospective universal tools instead, mathematics and common sense became alien to each other, even contradictory in their statements and, consequently, common sense was blamed for inferiority. (Keitel, 1996: 15)

Throughout my career in mathematics education I have encountered misconceptions concerning the nature of mathematics and people's perception of it and the way they react to and engage with it. Most people, for example, are surprised to learn that research mathematicians do not spend their time performing long calculations to several places of decimals. To say that mathematics is really about finding structure and logic and connections that help us negotiate the complex world we live in is like speaking an alien language for a vast majority of people. Yet, it does not take a great deal of effort, on an individual basis, to awaken most people to the power and fascination of mathematics through demonstrating practical applications or its beauty and elegance. So why is it that the criticisms about standards and dissatisfaction with the mathematical achievement of young people persist over centuries and have become almost institutionalised? The outcomes of, and responses of many countries, to TIMSS is a good example of this preoccupation with falling standards in mathematics.

It is now over twenty years since the Cockcroft Report was published and since then in the UK, an unprecedented amount of money and human resources have been spent on improving standards in school mathematics. In spite of it, it is still not difficult to find classrooms where children
- lack confidence in mathematics;
- spend the majority of their time reproducing their teachers' examples with different numbers;
- answer only other people's questions;

U. Gellert, E. Jablonka (eds.), Mathematisation – Demathematisation: Social, Philosophical and Educational Ramifications, 141–160. © 2007 Sense Publishers. All rights reserved.

- ask, ' What am I supposed be doing then?';
- fail to connect their mathematics with other subjects or with their life outside school, even when they are 'successful' in their mathematics classrooms;
- dislike mathematics, seeing it as irrelevant and boring;
- spend most of their time mystified.

In a study of 215, 11 years old pupils in an 8-12 years middle school, Haylock (1986) asked teachers to consider a list of statements which referred to various factors often associated with low attainment in mathematics. For each child with a score of below 20% on a standardised mathematics test the teachers were asked to indicate whether in their judgement, the statements described the child. These statements were based on previous studies such as Denvir, Stoltz and Brown (1982). Twenty-two statements are listed below with the percentage of low-attaining pupils for whom their teachers thought that the statements definitely or probably described them (cf., Haylock, 1986).

- has been considered low-attaining in mathematics from the first year in this school (82%)
- is low-attaining in most areas of the curriculum (79%)
- has poorly developed reading skills (77%)
- is equally poor in all aspects of mathematics (74%)
- has poorly developed language skills (70%)
- shows perceptual difficulties such as reversal of figures or poor spatial discrimination (45%)
- has immature motor skills (44%)
- is immature in relationships with other pupils (39%)
- shows little commitment or interest in mathematics lessons (39%)
- shows little commitment and interest in school in general (33%)
- has difficulty in relating to adults (33%)
- is nearly always preoccupied, appearing to find school and learning irrelevant (30%)
- has emotional problems related to an exceptional home background (30%)
- experiences social difficulties with the peer group (29%)
- displays behaviour problems, such as hyperactivity, in most lessons (26%)
- shows an abnormal level of anxiety towards most tasks in school (26%)
- shows an abnormal level of anxiety towards mathematics (26%)
- responds sensibly in a one-to-one conversation with a teacher but behaves badly in front of other children (24%)
- some physical factor such as deafness, poor eye sight, colour blindness, contributes to their low attainment in mathematics (18%)
- has been absent frequently in the last year (17%)
- seems excessively tired much of the time (14%)
- has suffered frequent changes of mathematics teachers or schools (5%)

The items at the top of the list suggest that pupils can be identified as low-attainers in mathematics by the age of 8, if not earlier and often remain in this category. Hart (1981) showed in her research that this often continued until the age of 16 when pupils left the school. *Table 1* below illustrates performance on three questions, clearly indicating little progress from age 12 to 15 years. In some cases, performance declined.

Table 1. Percentages of successful performance

Routine skills	Age: 12	Age: 13	Age: 14	Age: 15
263 + 978	85	89	88	88
2312 – 547	61	61	62	66
11/5 – 3/5	58	52	48	46

In contrast, consider my observation of 13 year old pupils who could not do simple algebra in the classroom but during the lunch time computer club they were using the most complex algebraic expressions. When I asked them to explain the algebraic notation and relationships, they managed to do it fluently and with confidence. This kind of experience occurs frequently whenever I have observed pupils and had conversations with them.

The point I am making is that from my observation, teachers and parents, who are the first point of contact with pupils, do not find it easy to identify the learning difficulties of pupils. Often, their own perceptions of learning mathematics influence them. They frequently describe the difficulties in general terms such as in Haylock's earlier items or by statements such as, 'they find fractions difficult'. Whereas in reality the nature of difficulty experienced by pupils can be varied and complex. Some of our current projects are concerned with sharpening this identification process by teachers.

I would like to illustrate the above further by using some examples. Take 11 year old Charlie, for example. When asked to subtract 70 from 109 on paper, he could not do it. His explanation went: "Zero from 9 you cannot do so put zero down. You can't take from 7, so put zero. There is nothing to take from 1, so you put 1". His answer therefore was 100. However, when asked to take £70 from £109, he immediately worked out by adding on to £70 that it would take £39 to get £109.

What this episode demonstrates is the gap between how people cope with mathematical situations they meet outside the classroom and formal classroom approaches. When teachers have seen the video of Charlie, their immediate reaction is to suggest that practical ideas and models should be used to explain the subtraction question. This can be as often counterproductive as helpful. They have rarely considered that the simple subtraction sign, used in presenting the problem to Charlie, could be hiding many meanings behind it. For example, see *Table 2*, taken from Ahmed and Williams (2002), below.

In the UK, there seems to be a great tendency among teachers to look for good ideas and resources for classrooms. If they work, they will be used again. If not, then they can always look for or wait until they find other resources. Among the aids used, blocks and tiles are often used to teach topics such as fractions which pupils find difficult.

Table 2. 14 –9?

What is the difference between 14 and 9?	How much bigger is 14 than 9?
	14 minus 9
Subtract 9 from 14	From 14 subtract 9
What would you add to 9 to make 14?	How much more than 9 is 14?
If you have 9, how many more do I need to make 14?	What is 9 subtracted from 14?
	Take 9 from 14
	14 take away 9

I would like to probe the use of such models further. Consider the use of fraction blocks to help pupils add 1/2 + 1/4.

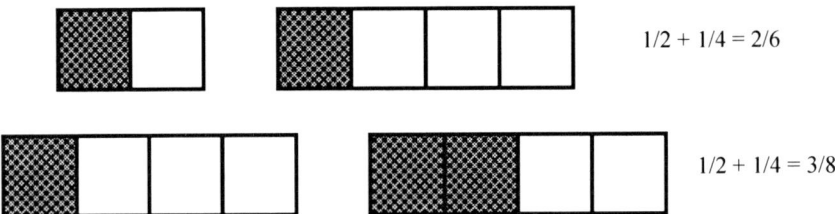

$$1/2 + 1/4 = 2/6$$

$$1/2 + 1/4 = 3/8$$

Figure 1. 1/2 + 1/4 = ?

Some pupils drew two rectangles and divided one in halves and another in quarters and counted the shaded parts to get the answer 2/6 and the total number of parts, others divided the two rectangles in quarters and did the same to get the answer 3/8. It was not even clear if those pupils who got the correct answer of 3/4 could justify why the answer was correct with reference to the tool used. Obviously, the use of didactic materials/tools can play an important role in the discovery and expression of relationships. However, it is not a substitute for teachers enabling children to articulate and define their understanding of mathematics represented by the tools they used.

The interplay among and connections between objects (structured or unstructured), images, language and symbols that lead to mathematical reasoning and the stating of mathematical propositions of very wide generality is well worth a closer study. I believe that the subtle distinction between the way mathematical ideas are constructed from objects and the particular characteristics of the objects is often not clear in many teachers' minds. For example, when we draw a triangle on a sheet of paper and by means of this we prove a general proposition that the three angles add up to 180°, it is worth reflecting how it is that this very particular triangle leads us to deduce something of such wide generality. If we examine closely the figure we have drawn, it will be obvious that it is not a triangle at all – three rather uneven marks, possibly with blunt corners! This does not seem to make any difference to the proof, though. In this case the triangle really is an idea, not an object. It is a mental image drawn from the real world, which aids mathematical thinking and can be much richer than an object. Papert (1980) described such mental vehicles as "objects-to-think-with."

A considerable proportion of my work and, indeed, the central tenet of the Mathematics Centre's work at my University is that mathematics is more accessible to more people than it is believed generally and our efforts have been directed at achieving this.

The problem with mathematics is that it has had a very long past as a school subject and an academic discipline and has no reason to establish any credentials – it is universally accepted as an essential core academic and useful discipline and any attempts to make it less important will meet with violent resistance. For example, I remember that in the 1970s, a number of Inner London schools made mathematics optional for pupils in years 10 and 11 (15 and 16 years old). Hardly any pupils dropped the subject, although most of them loathed it and anyway there were no mathematics teachers to teach them. This may offer us a clue to why it is so very difficult to make mathematics a more appealing subject in schools.

My first proposal is that the very respectability of mathematics and almost unthinking acceptance of its importance by a whole sector of society prevents a sensible discussion leading to productive actions in alleviating concerns about standards in the subject – from primary to post-graduate level. For example, for many, the success or failure of research and curriculum development projects can be decided on whether there is any significant improvement in the test scores in mathematics – irrespective of whether any aspect of mathematics or numeracy was part of the overall project design. This is reinforced by a popular notion that mathematics is *value free*, can be tested quickly and would render *pure* evidence of achievement. In addition the following two aspects have particular significance in discussions:

– the elevated status of mathematics in society produces a tendency to over-rate the information yielded by test and examination results and under-rate other measures. The belief in the more evident 'objectivity' of measurable elements has hindered progress in introducing more effective procedures for acknowledging mathematical achivments. In practice, the various forms of assessment and evaluation need to complement one another (fore a more fully discussion of this issue, see Ahmed (1987));

– the view of mathematics which seems to be interpreted as a body of established knowledge and procedures – facts and rules. This describes the form in which we observe mathematics in calculations, proofs and standard methods. However, most mathematicians would see this to be a very narrow view of the subject. It is certainly the case that in many school mathematics lessons, pupils are taught rules, technical notation and established conventions without acquiring any feeling for why these systems exist.

Mathematics is hard because it deals with abstract relationships, but some parts of it look easy because it deals with symbol-manipulations, which it is possible to learn without understanding the underlying operations to which they refer. (Bell, Costello, & Kuchemann, 1983)

Commenting on the treatment of school mathematics curricula in international comparative investigations, Keitel (2000), wrote,

... international comparative studies have come to dominate educational discourse in many countries when educators, policy makers and politicians consider what is to be done to improve their system of education. The results of these studies are accepted in many places as providing undisputed scientific evidence about the achievement of students as well as indicating how good the curriculum is and how well teachers are teaching. (Keitel, 2000: 43)

... the idealised international curriculum defined by a common set of performance tasks organised by content topic remains the standard of measurement. No allowance is made for different aims, issues, history, and context across the mathematics curricula of the system being studied. (Keitel, 2000: 44)

The above identifies a further reason why mathematics is viewed by many as a set of procedures to be practised and memorised. At a first glance this could appear relatively straightforward to attempt to address but in reality the situation is complex.

There is a wide range of perceptions concerning the nature and definitions of mathematics and its place in school curriculum. Some of these definitions are matters of philosophy, some arise from the perceived needs of industry, science and other disciplines. Most of them have historical origins often derived from social contexts; indeed most people's perceptions of mathematics are shaped by the way they were taught at school.

There are a high proportion of people, in all walks of life, who use mathematics effectively without realising that they are doing so and yet undervalue their own mathematical confidence and competence. Hence their mathematical growth has been stunted and their potential has not been realised. It is not surprising to find that people who cannot do mathematics revere those who think they can (Sewell, 1982). The mad mathematics professor's image of 'being too clever to be understood' lives on in people's minds, colouring their judgement and actions. Often the elevated status of mathematics in society is attained through its undermining effect on many adults and children. Perhaps it is obvious that these are the very people who place achievement in mathematics as a most important attribute. A range of surveys and studies has exemplified this notion. For example in the late 1980s, the *Sunday Observer* newspaper recorded the responses of a survey carried out to determine which three school subjects were considered by the public to be most and least important.

Table 3. Importance of school subjects

Most important		Least important
81%	Mathematics	Less than 0.5%
79%	English	1%
29%	Computer Studies	4%
22%	Science	4%
15%	Foreign Language	17%
10%	Business Studies	10%
8%	Technical Subjects	5%
8%	Religious Education	37%
7%	Sex Education	28%
6%	Home Economics	16%
6%	Social Studies	19%
6%	Economics	6%
6%	History	14%
6%	Physical Education	18%
3%	Geography	8%
3%	Woodwork/Metalwork	12%
2%	Peace Studies	49%
1%	Art	35%

We have been associated with similar surveys in parts of Australia and New Zealand which reflect the same trend, particularly in the rating received by mathematics. I am in no doubt that a similar response will be obtained if the survey was undertaken today.

I will now consider an example of how mathematics can be used (or abused) in bringing about respectability in another discipline, for example, Chemistry. The subject must constitute one of the world's largest bodies of accumulated knowledge. To bring order to this data, theories are needed. The ultimate goal is towards a grand unifying theory that explains all chemistry. Mathematics is the key most scientists rely on to describe everything in the physical world. Since it works spectacularly well, it is seen as an unambiguous business that leaves no scope for subjective contributions or individuality of choice.

In 1926 Erwin Schrödinger discovered his famous wave equation. This gave the chemists their super theory and with it came respectability. Implicit in Schrödinger's wave equation is all the information about atoms and molecules that a chemist might wish to know. Unfortunately the equation cannot be solved for any but a handful of trivial molecules, yet such a store was set upon it that undergraduates were often greeted with it in their first chemistry lectures at university. Curiously, it is one of the least useful concepts that students need for their subsequent career.

147

A growing body of opinion began to question the value of the wave equation in teaching chemistry although there was no doubt about the usefulness of theoretical chemistry in basic research. Among this lobby, people such as Robert Sanderson of Arizona State University attacked the notion by saying that the chemists were 'brainwashed into believing that Schrödinger's equation is the complete guide to chemical understanding'. The lobbyists who advocated descriptive chemistry were not rejecting the wave equation but were pleading for concentration, at first, on information that was useful for a chemist rather than what fits mathematical logic best. They wanted, 'concepts rather than calculus, functional groups rather than group theory'. They wanted descriptive chemistry to capture the excitement of making something new, of understanding why compounds behave as they do, and of using mathematics to probe the real world; not to construct a perfect but imaginary world.

There are many parallels with the debate on teaching applied mathematics in the UK. One has only to see the range of books designed to help students to solve *real-world problems*. In most cases the models are too complex to allow for much to be learnt from a mathematical analysis. So various simplifications are introduced to arrive at neat mathematical equations. The focus then becomes fluency with particular mathematical techniques and the real problem is still far from being solved. It is, therefore, not surprising that students have little experience of the power of mathematics in solving practical problems.

As signalled earlier, I believe that the problems of learning and teaching mathematics are associated mainly with people's belief about what constitutes mathematics, how it is learnt and applied, how it is taught and how best to assess this.

A narrow, techniques and conventions orientated view of mathematics is inconsistent with descriptions of the skills needed for the 21st century and espoused by governments and industry, internationally. At a pragmatic level, technological changes, especially in the means of production and communication have raised the demand for skills level needed for employment and transformed the structure and management of organisations. Hence the users of mathematics cannot afford to solely identify mathematics with effective techniques without being conscious of the rigour of mathematics which guarantees the validity of the techniques used. Similarly it may be said that the academic mathematicians need to be sufficiently informed about applications of mathematics in business, industry, other areas of education and everyday life to see their own work from user's viewpoint, and to value it for the practical applications as well as for the academic scholarship.

The emphasis has moved to outcomes such as 'transferable' skills, adaptability and positive attitudes to change, the ability to communicate, desirable personal and interpersonal qualities and problem solving capabilities. Countries throughout the world tend to want to move towards a more vocational approach to general education and a more general approach to vocational education. It also appears to be acknowledged that focusing on specific skills will not achieve the desired outcomes such as practical capability and the ability to use knowledge and skills as resources in solving real-world problems. Of necessity, one would assume that aims and definitions of mathematics within school as well as higher education curricula will need to move from those set up fundamentally to achieve a narrow range of competencies to a situation where students will need to achieve a broad range of skills inclusive of those which have been in the past sufficient to address needs.

But what priorities should a mathematics curriculum address within individual country contexts? Obviously a curriculum cannot be value and culture free. Howson (1991) indicates that:

> It must nurture and challenge those who are to provide the country with intellectual, technical, industrial and commercial leadership. There is, therefore, an elite for which the system must cater; and it's now accepted in most countries that entry to the elite must be open to all – it cannot be reserved for males, or those belonging to certain social classes or ethnic groups or living in particular geographical areas.

Apart from the elite, mathematics is increasingly becoming an important tool for technical and commercial studies as well as aiding people's ability to comprehend the world around them and helping them to become worthwhile democratic citizens. For example, the place of mathematics as a core subject in the National Curriculum of England and Wales is described as follows in the Non-statutory Guidance (1989).

> Mathematics provides a way of viewing and making sense of the world. It is used to analyse and communicate information and ideas and to tackle a range of practical tasks and real-life problems.

> Mathematics also provides the material and means of creating new and imaginative worlds to explore. Through exploration within mathematics itself, new mathematics is created and current ideas modified and extended.

> Both in tackling problems and in exploring within the subject itself, mathematics has the capacity not just to describe and explain but also to predict. This gives mathematics the power and pervasiveness that accounts for its importance in the school curriculum.

At first glance, the above could appear relatively straightforward to attempt to address by introduction of new syllabuses, modes of examination, resources, organisational structures, teaching styles, training and expert demonstration. In reality the situation is complex. What might appear to be clear when expressed in words, might and assume different conceptual frameworks other than those in the minds of those who conceptual and write. Beliefs about the nature of mathematics have a direct impact on what is taught, and how it is taught. Students, teachers, mathematicians, textbook authors, parents, employers and politicians – all exercise their influence on educational practice and transmit an element of the beliefs they hold to the mathematics classroom. It is not easy to disentangle the different influences, let alone identify the extent to which each affects the outcome.

So if mathematics is increasingly important for all and we want to promote its accessibility as well as enabling the *elite* to push the frontiers of the mathematical world, what are the ways forward?

I would argue that we cannot make progress unless *all* those who have influence on teaching and learning mathematics at all levels understand its power and limitations. Stigler and Hiebert (1999) propose that teaching is a cultural activity and the improvement process should begin by becoming more aware of cultural scripts teachers are using.

> This requires comparing scripts, seeing that other scripts are possible, and noticing things about our own scripts that we had never seen before. […] No matter how good the teachers are, they will be only as effective as the scripts they are using. To improve teaching over the long run, we must improve the script.

From my point of view the *cultural scripts* teachers use in connection with the learning and teaching of mathematics are subject to a strong influence from the diverse groups, outlined above, which impact their beliefs. Hence changing their *scripts* is also important but a daunting task.

We might even need to confront the big questions such as: *Is mathematics something we discover, or do we invent?* Obviously such a question is an inappropriate one to consider here but I would like to explore some of its implications.

In a BBC television series, *A Way with Number*, designed to help adults who failed to understand mathematics at school, there was an interesting visual demonstration of taking a square and continuing to cut it in half in order to generate fractions which all add up to one whole. A simple practical situation such as this can easily lead to abstract notions of infinite series. One has only to ask questions such as, *What would happen if ...? How, When?*, in order to move from a practical to an abstract situation which could hold an equal appeal for many, similar to the appeal Su Doku holds for many.

I consider that, particularly in the current technological context, the dichotomy between teaching *mathematics as mathematics* and *mathematics as a means to an end* is an irrelevant one. The areas of application of mathematics are far less precisely defined than previously, when there was a consensus about the mathematics the traditional scientists needed. We have many users of mathematics – geographers, urban planners, industrial economists, demographers, medical diagnosticians, to name but a few. To meet such a range of diverse needs, it is unrealistic for students of mathematics to spend a large part of their time acquiring specific skills, before it is known to them which ones will be needed – not to mention the effect on motivation. In addition, the knowledge explosion will not allow us any more time to add to already over-crowded syllabuses.

What about the increasing need of mathematics as a means towards being a more numerate and hence effective citizen?

The solutions to this will have to have a different rationale to that adopted in response to C. P. Snow's 1959 lecture, *The Two Cultures and the Scientific*

Revolution. Many of the solutions proposed were in terms of making it compulsory for students to study one or two subjects from the opposite side of the 'Snow Line' to that in which they wished to specialise, hence offering them a rounded educational base.

In Roger Penrose's book, *Shadows of the Mind* (1994), he writes about three worlds and three mysteries. His three worlds are:
− Mental world
− Physical world
− Platonic world

His mental world is what he calls *the world of our conscious perceptions* and it is a world that we know least about. It includes pain, happiness, perceptions of colour, fear of death, knowledge and understanding of numerous facts, as well as ignorance and revenge. It is a world where smells, sounds and sensations of all kinds intermingle. It also contains mental images of chairs and tables.

His physical world contains actual chairs and tables, human beings, human brains. It also includes clouds, hurricanes, rocks, flowers; at a deeper level are atoms, electrons and space-time.

His platonic world is a world of *mathematical forms*. This world contains natural numbers and the algebra of complex numbers, Euclidean and non-Euclidean geometry, infinite and non-computable numbers, recursive and non-recursive ordinals. It contains Turing machine actions that never come to an end. It also contains mathematical simulation of chairs and tables, as would be made use of in virtual reality, and simulations of hurricanes and black holes. This is the world whose existence rests on the timeless and universal nature of concepts and laws which are independent of those who discover them.

Without re-rehearsing the whole of Penrose's argument about the relationship among the three worlds, it would suffice to point out one fact. He has drawn his diagram to illustrate, seemingly, the paradox of how he sees each world emerging from only a tiny part of the one which precedes it. He is not indicating, which, if any, of the worlds are to be regarded as primary, secondary or tertiary.

I have referred to Penrose both to illustrate how little we understand human thinking but also to indicate new and exciting insights into mathematical thinking. It also highlights, for me, the nonsense of any individual or agency answering the plea for raising mathematical standards with a stable syllabus, teaching methods and assessment system. It would also be erroneous to view separately the theoretical, applied and practical aspects of the learning and teaching of mathematics. Most of all we have to be as open to learning from new insights on the nature of mathematics and human mathematical cognition as we have to be about new theories of learning and prepared to continually review and change our practices accordingly.

At this stage I would now like to propose my perceptions of the three worlds in order to help demystify polarities between 'real' and 'abstract' mathematics which often tends to hinder students and teachers in their attempts to learn and teach mathematics.

My worlds are

- *World 1*   *Common sense.* This is concerned with the large proportion of people of all ages who use mathematics (both process and content) effectively but do not call it mathematics. The Cockcroft Report (1982) described this phenomenon as 'If I can do it is common sense; if I can't, it is mathematics'.
- *World 2*   *Mathematics as a means to an end.* This is concerned with a world where people consciously learn some mathematics because it is necessary in order to solve a problem. The problem could be concerned with building a pond or understanding another subject area.
- *World 3*   *Mathematics for its own sake.* This is concerned with a world where people enjoy the appeal and beauty of mathematics for its own sake, including the group I might irreverently label 'theorem proving mathematicians'.

I consider that it is entrenchment in these three worlds that is at the root of the problems concerned with underachievement in mathematics. At a first glance my first world and my third world would seem poles apart. From my experience, the reality is that they are not necessarily so, but keeping them apart for pupils makes it so.

It is only when students and teachers have experienced, appreciated and understood a wider view of mathematics that we will begin to develop a mathematically confident society which can bridge the three worlds, I described above, and enable people to excel in each of the three worlds according to their own priority and attitude.

As part of the Low Attainers in Mathematics Project – LAMP[53], some teacher-researchers found it useful to build on pupils' natural activities outside the classroom to enable them to develop familiarity with mathematical processes such as classifying, symbolising, proving, transforming, and conjecturing.

Table 4. Activities in which children naturally engage

Experiment	pull apart, find how things work, find how people react, testing things/ people, speculate, challenge, change things – break rules, make things
Organise	collect, separate, spend, save, budget, time
Create own world	control over environment, codes, language, symbols, jokes, fashion
Compete	
Self-evaluate	
Imagine – Fantasise	make up stories, invent
Create and solve mysteries	
Think about adult life	jobs, earning money, leaving home, leaving school
Develop independence	

Communicate	explaining, arguing, asking, expressing opinions, sign language, symbols
Use leisure	watch television, sports, reading, art/craft, technology
Repeat and practice	sports, learning to play musical instrument

My colleague, Marion Bird's work with young children 3 to 11 and the work arising from the Raising Achievement in Mathematics Project (RAMP), successor of LAMP, has demonstrated how links between the first and third worlds can be made in the context of schools, leading to increased access and achievement for pupils (see for example, Bird (1991)).

My second world, concerned with the application of mathematics, can benefit most when there is a close link with world 3, i.e. using rules with understanding. Through my example of cutting a square into fractional sections, I have already indicated how world 3 is linked with world 1. Obviously at a school level the aptitude and interest of pupils and the context in which they learn will dictate the levels of their achievement within worlds 2 and 3.

The interaction among these three worlds of mathematics will be manifested differently, for example in developed and developing countries but the commonality will be in curriculum planning, the emphasis on how people can learn mathematics effectively and the necessary pre- and in-service provision for teachers. No curriculum will achieve its objectives if the participants (mainly teachers) do not fully understand both the changes that may be required and an appreciation of the nature of mathematics itself. As an example, research on the development of universities in developing countries (Lusthaus, Anderson, & Adrien, 1989) indicated that millions of dollars were being spent to develop universities without a clear understanding of what is meant by university development. Similarly, unless teachers discuss, debate and consider the implications of questions such as, '*Does the content of mathematics courses taught provide a valid impression of mathematics as it exists today both in terms of major ideas and the evidence upon which those ideas are based?*', the outcomes of their practice will be limited by their view of mathematics and hence limit their pupils' horizons about mathematics and its purposes. One of the conclusions of the RAMP project was that facts and skills taught in isolation from conceptual structures and general strategies can undermine pupils' confidence and competence. From our experience, particularly in working with primary teachers on in-service courses, the introduction of the National Curriculum in England and Wales (1988-) and the National Numeracy Strategy in recent years, led teachers to concentrate mainly on planning and on improving teaching approaches and a second priority being given to the study of how pupils form elementary mathematical concepts. What is a contemporary issue for effective teaching and learning as identified from research studies may not be the key issue in the school staff room.

How influential has the literature been in reflecting policy into practice?

In order for research findings to become useful, inevitably teachers will need to set their own boundaries and create a framework to incorporate these findings. It is impossible to maintain a clear division between the literature on teaching from those of learning i.e. to separate issues of curriculum and pedagogy from the contexts within which the curriculum operates. Given the above, it was necessary to bring about a clearer understanding of the ways in which teachers can effectively plan and implement the teaching of mathematics. In other words, to move from what could be traditionally be thought of as a 'Tyler' model (1949) to one which mirrors a model developed for science (Hodson, 1993) and which we have developed for mathematics teaching and learning (Ahmed & Williams, 1995).

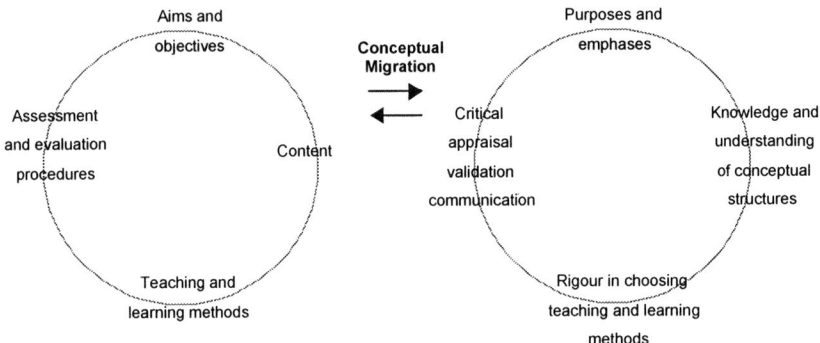

Figure 1. A model for effective planning and implementation

The concept of *migration* in the above model is important since it offers a slow but more positive way forward for changing practice than highly centralised approaches governments such as ours have become used to adopting in order to effect a quick change. The model above enables teachers to become progressively reflective without becoming disillusioned with the day to day reality they face in their institutions.

The first issue of *The Bulletin of The Association for Teaching Aids in Mathematics* (November 1955) outlined in the Editorial that it will carry articles covering the whole field of mathematics teaching, paying particular attention to the use and development of teaching apparatus and visual aids. It went on to say,

> the teaching of mathematics requires constant research; and research which aims to advance knowledge of the craft of teaching is just as difficult as research which aims to advance knowledge of mathematical techniques, and perhaps it is even more important. No one can do it better than those who are actively working in the classroom [...].

In order to illustrate our approach to the teacher involvement in their own research, I will briefly describe a project involving both primary and secondary phases of schooling (5-14) years (Ahmed & Williams, 1997). The project's focus was on teacher development in order to improve pupils' numeracy with the underlying premise that teachers needed to be aware of and gain understanding of relevant research in the area and become involved in research themselves.

The project was designed to respond to this context with a premise that teachers are key agents of implementation in the classroom. It involved two cohorts of 40 teachers from 53 schools, representing all four regions of the local education authority and across key stages 1, 2 and 3. The participating teachers reflected a range of background experiences and viewpoints across the key stages and also included headteachers and mathematics coordinators. During the first phase, the teachers met with The Mathematics Centre's advisory staff on a regular basis to clarify and interpret competencies in numeracy in their own context, share and reflect on classroom experiences, plan individual and shared classroom tasks and analyse classroom interactions.

The project's main limitations were concerned with the availability of resources. The formal initial phase constituted, for teachers, one full day and three twilight sessions followed by one day to work in other classrooms each term. The rest of the classroom-based research was carried out during teachers' normal teaching time. The major source of school-based support was offered by the associate adviser who, in addition to visiting schools, ran area-based clinics for teachers wishing to discuss their work. The project team met frequently to review progress and plan strategies.

As the project progressed, informal contacts took place among teachers sharing similar interests. The research process constituted
- the development and documentation of teaching strategies and learning resources which contribute to raising achievement;
- exploration, interpretation and documentation of the nature of pupil and teacher support which leads to successful teaching and learning approaches in the area of numeracy;
- the use, relevance and benefits of using resources (including technology) appropriately and how to become more discerning users of existing published materials.

In schools, teachers explored possibilities and approaches in their own and colleagues' classrooms. School visits were made by the advisory staff to support teachers and by The Centre staff to aid the research process and evaluate outcomes. A key element in the school-based work was the requirement for teachers to work in classrooms with a colleague from another school and from a different age range.

A case study approach was adopted. A case study is an empirical inquiry that (a) investigates a contemporary phenomenon within its real life context when (b) the boundaries between phenomenon and context are not clearly evident and in which (c) multiple sources of evidence are used.

All teachers kept logs and wrote case studies of their work as they needed to monitor and evaluate their work so that effective strategies could be formulated at every stage.

The logs and case studies were analysed by the team from three perspectives, in terms of:
- substantiation and justification of findings;

- evidence of pupil achievement and teacher achievement;
- usefulness to other teachers within and outside the project.

Particular trends or categories emerging from the logs formed a basis for checking the logs again. These were confirmed further by follow up visits to schools to work with teachers, observing lessons and asking further questions for clarification of the process.

The extent to which teachers became engaged in the project is as complex as the project's focus. For some teachers the project was a means of improving their classroom practice, others registered for a research degree or other postgraduate qualifications as a means of offering a focus and discipline for their work.

Some extracts from teachers' field notes will help illustrate the process:

Extract 1

Not surprisingly I have not managed to answer conclusively any of the questions posed at the beginning of my work. In carrying out this research I have become aware of certain difficulties and restrictions. It is very difficult to be unbiased and neutral when working with a group of children whom you know well and when you have an hypothesis that you would like to prove. As I have said previously, ideally I would have liked to record the Year 7 [12 years old] in action but the restrictions of a large group, limited space and background noise makes it impossible. Making a written record as it took place was possible for short periods of time.

Extract 2

From my field notes

Andrew   January

Andrew was making much use of his fingers to count on and back, not particularly successfully. For example
$93 + 8 = 141$ (he couldn't see that his answer was too high)
$34 - 20 = 3$ (he counted back and again didn't see his answer as unreasonable)
$27 + 50 = 77$ (this took him four attempts, he counted on from 50, but not in tens)
He had more success with numbers which totalled less than 10, for example,
$10 - 8 + 2$ "because I know 2 and 8 is 10"
$32 + 5 = 37$ "because 2 and 5 is 7"
He wouldn't try $420 + 390$ because he said he couldn't do it.

Andrew   March

Andrew has become more adventurous and more accurate and is using his fingers less, for example
$16 - 4 = 12$ "I just knew it"
$56 + 34$ "50 and 30 is 80, 6 and 4 is .... Oh the answer in 90"

Whereas earlier in the term Andrew wouldn't try 420 + 390 he was happy to attempt 710 + 160 this time,

"700 + 100 is 800, 10 and 60 is 70, add it up – 870"

With 37 + 58 he initially split the tens and units, added 30 + 50 to make 80, then added 8, and lastly counted on 7 to arrive at the answer 96. The rest of the group told him he was wrong and I asked him if he could think of another way of doing it. He said he could round it up so that 58 became 60 and round down the 37 to 35, this then became 60 + 30 + 5 =95

Implications for future teaching

The most obvious improvement has been that the children are more confident in their approach. They are attempting more difficult calculations, even if they weren't always getting them right the first time. It was pleasing to see that they were making use of the mental strategies I had covered in class lessons.

However I am concerned about the tendency to automatically split numbers into tens and units or hundreds, tens and units. This concern has been reinforced by my experience with small groups of children in years 4 and 5 who have learnt the pencil and paper method for addition and subtraction and now seem almost 'locked in' to using it in every situation.

I need to continue to discuss with the children on a regular basis the range of strategies that can be used in mental calculations; they need to practise these strategies but they also need practice in choosing the most appropriate strategies. One way of practising this might be to give the whole sum with the answer, for example, 65 + 99 = 194 and ask the children for a sensible method of getting the answer.

I shall be encouraging the children to count on more sensibly and I will be dissuading them from splitting numbers into hundreds, tens and units at least until they have considered other strategies.

Extract 3

In the past two years my own views of mathematics, how people learn it, and how it should be taught have changed. The stimulus for these changes has been a combination of my own experiences of doing mathematics and the discovery that my pupils could do mathematics in ways that I had not appreciated before. As I have altered the way I teach mathematics, I have found pupils have been more highly motivated and have demonstrated skills that I had not suspected they possessed.

Extract 4

Throughout this time I have become increasingly aware of the importance of the quality of teacher-pupil and pupil-pupil interaction and have therefore chosen this area for my further research. Before undertaking this study I needed to consider the following:

When should observations take place?

Where should observations take place?

Should the focus be on the teacher, teacher-pupil interaction, pupil-pupil interaction, some or all of these?

What data recording methods should be adopted?

What aspects of interaction should be studied?

How will the data gathered be used?

I looked at Flanders Interaction Analysis Categories and although it described a number of behaviours that I believe are important in the classroom, non-verbal aspects of interaction were lost and some categories were too broad. I began to devise my own interaction analysis categories and found difficulties in deciding on categories concerned with children's responses. After abandoning correct/incorrect, acceptable/unacceptable and direct/indirect I have used appropriate/inappropriate to describe their responses. I believe that my categories, if used by teachers in their own classroom research, would enable them to get a reasonably full picture of classroom interactions. I also selected the following learning skills which I believe are relevant to my work. These are:

Do pupils have a positive attitude to work?

Do they remain on task?

Are pupils able to work independently; are they able to work collaboratively?

Do pupils take responsibility for their own learning?

Finally, I decided to look closely at the mathematical language used in my case studies.

I used a video camera in all observations, thus allowing me to keep a permanent record of collected data. I was able to replay and re-check points discussing these with my colleagues, to focus on events and to carry out analyses of my own actions and re-actions. I could look at non-verbal features such as body language, facial expression and silence.

I also scribed a number of interesting interactions as, I have discovered in the past, even in the quietest room dialogue can be missed using a video camera.

The personal development of teachers was the most important factor in helping raise their students' achievement. By necessity this involved them in examining critically their beliefs and practices and comparing their findings with previous relevant research and curriculum studies as well as with other teachers' findings. Both as researchers in their classroom and critics of other research, their ability to facilitate their colleagues' development and their students' learning increased considerably. The inability of some teachers to share experiences with others was the greatest inhibitor to progress.

Once again, I will bring Christine Keitel to my aid in helping me conclude these explorations:

The domain of mathematics education cannot be seen as controlled by culture-free laws which are to be progressively identified through research. Also, many big theories and concepts or metaphors used to support them, are simplistic and have received more attention than they deserve. The idea that the best mathematics education research is that which is based on a coherent theoretical framework has to be questioned. Mathematics education needs more reflective, culture-sensitive, and practice-orientated research. (Keitel, 1998: viii)

NOTES

[53] The Low Attainers in Mathematics Project (LAMP) was set up in response to the Cockcroft Report (1982) and was charged, in the first instance, with the task of developing good practice in the teaching of mathematics to low attaining pupils and then to investigate means of disseminating its findings.

REFERENCES

Ahmed, A. (Project Director) (1987). *Better mathematics*. London: HMSO.

Ahmed, A., & Williams, H. (1992). *Raising achievement in mathematics*. London: WSIHE/DES.

Ahmed, A., & Williams, H. (1995). Key features underpinning the effective teaching of numeracy, unpublished research paper, university college of Chichester, Chichester.

Ahmed, A., & Williams, H. (1997). Numeracy project: A catalyst for teacher development and teachers researching. *Teacher Development, 1*(3), 357-373.

Ahmed, A., & Williams, H. (2002) *KS2 mathematics: Numeracy activities: Plenary, practical & problem solving*. Network Educational Press.

Association of Teaching Aids in Mathematics (1955). Editorial. *The Bulletin of The Association for Teaching Aids in Mathematics, 1*, 2.

Bell, A. W., Costello, J., & Kuchemann, D. (1983). *A review of research in mathematical education: Part A, Research on learning and teaching mathematics*. London: nferNelson.

Bird, M. H. (1991). *Mathematics for young children*. London: Routledge.

Cockcroft, W. H. (Chair) (1982). *Mathematics counts: Report of the Committee of Enquiry*. London: HMSO.

Denvir, B., Stoltz, C., & Brown, M. (1982). Low attainers in mathematics, Schools Council Working Paper 72, Methuen.

Hart, K. M. (1981). *Children's understanding of mathematics: 11-16*. London: John Murray.

Haylock, D. (1986). Mathematical low attainers checklist. *British Journal of Educational Psychology, 56*, 203-208.

Hodson, D. (1993). Teaching and teaming about science: Considerations in the philosophy and sociology of science. In D. Edwards, E. Scanlon & D. West (Eds.), *Teaching, learning and assessment in science education*. Milton Keynes: The Open University/Paul Chapman Publishing.

Howson, A. G. (1991). *National curricula in mathematics*. London: The Mathematical Association.

Keitel, C. (1996). Discussion paper: Mathematics (education) and common sense. In C. Keitel, U. Gellert, E. Jablonka & M. Müller (Eds.), *Mathematics (education) and common sense: The challenge of social change and technological development*. Berlin: Freie Universität Berlin, 14-28.

Keitel, C. (1998). An introduction. In C. Keitel (Ed.), *Social justice and mathematical education*. Berlin: Freie Universitat Berlin, v-viii.

Keitel, C. (2000). Cultural diversity, internalisation and globalisation: Challenges or perils for mathematics education? In A. Ahmed, J. Kraemer & H. Williams (Eds.), *Cultural diversity in mathematics (education)*. Chichester: Horwood, 41-57.

Lusthaus, C., Anderson, G., & Adrien, M. H. (1989). Changing trends in the role of Canadian university in development assistance. In C. Lusthaus & G. Anderson (Eds.), *Changing trends in Canadian development assistance*. Montreal: McGill-Queens.

Papert, S. (1980). *Mindstorms: Children, computers and powerful ideas*. New York: Basic Books.

Penrose, R. (1994). *Shadows of the mind*. Oxford: Oxford University Press.

Stigler, J. W., & Hiebert, J. (1999). *The teaching gap*. New York: The Free Press.

Sewell, B. (1982). *The use of mathematics by adults in daily life*. London: Advisory Council for Adult and Continuing Education.

Snow, C. P. (1959). *The two cultures and the scientific revolution*. Cambridge: Cambridge University Press.

Tyler, R. W. (1949). *Basic principles of curriculum and instruction*. Chicago: University of Chicago Press.

AFFILIATIONS

Afzal Ahmed
Mathematics Centre,
University of Chichester

JEREMY KILPATRICK

DEVELOPING COMMON SENSE IN
TEACHING MATHEMATICS

INTRODUCTION

In *Mathematics Education and Common Sense*, Keitel and Kilpatrick (2005)
observe that mathematics teaching too often privileges a study of the discipline
over an explicit development of learners' common sense. At the same time, the
hidden curriculum of mathematics teaching ends up creating a common sense for
learners that teachers seldom share. What is true for teachers and learners can also
hold for mathematicians and teachers. In this paper, I explore some examples from
recent disputes – the so-called math wars – in which the presumed commonsense
view of school mathematics being proposed by mathematicians is not seen as
common sense by teachers. One means of reaching common ground in the math
wars, therefore, might be for mathematicians and teachers alike to educate their
common sense.

DIFFERENT VIEWS OF COMMON SENSE AND SCHOOL MATHEMATICS

The mathematician George Pólya was fond of pointing out that although we can
usually rely on our intuition for ideas on how to solve a mathematics problem,
sometimes our intuition lets us down. We need to, in his words, "educate our
intuition". In mathematics education, intuition is often seen as akin to common
sense (Fischbein, 1987; Freudenthal, 1991), and our common sense needs to be
educated, too. Common sense is both an individual possession and a social
construction. It helps us learn, do, and teach mathematics, and it also can hinder all
those processes.

School mathematics has historically attempted to mirror what has been seen as
the abstract, context-free, universal nature of academic mathematics.
Consequently, mathematics teaching has tended to concentrate on the promotion of
skill in handling routine numerical, algebraic, and geometric operations divorced
from meaningful contexts or realistic applications. Far from drawing on, let alone
developing, learners' commonsense notions of quantity and space, instruction
seeks out the rarefied realm of abstraction, formalism, and generality. With few
exceptions, learners respond to such instruction with boredom and indifference.

The common sense developed by learners thereby becomes a perspective that is
less about mathematics than about how to cope in the mathematics classroom. A

*U. Gellert, E. Jablonka (eds.), Mathematisation – Demathematisation: Social, Philosophical and
Educational Ramifications, 161–169.* © *2007 Sense Publishers. All rights reserved.*

curriculum that is all but hidden from the teacher educates learners as to the minimum amount of effort the teacher will accept, what will count as evidence of interest and learning, answers the teacher expects to hear, and what ought not to be said or done. Each group then works within its own common sense, with teachers pitting mathematics against what they see as ignorance and common misconceptions, and learners pitting their wits against what they see as senseless activity. Teachers and learners are separated not merely by faulty communication but also by different views of common sense and school mathematics.

Two other groups that can be separated in a similar fashion are mathematicians and schoolteachers. The teaching of school mathematics involves the interactions among a teacher, some learners, and some mathematical subject matter in which they are jointly engaged. Mathematicians typically see themselves, but not teachers, as experts on subject matter. Teachers typically see themselves, but not mathematicians, as experts on pedagogy and learners. In each case, there are opportunities to educate what the group sees as common sense about its expertise and the other group's lack of expertise.

STANDARDS FOR SCHOOL MATHEMATICS

Some particularly rich examples of divergent perspectives on common sense and how they might be addressed can be found in controversies that developed in response to efforts by groups of mathematics teachers in North America to change school mathematics. Unlike the new mathematics movement of the 1950s to 1970s, which began with university mathematicians, the recent movement was initiated by a professional organisation, the National Council of Teachers of Mathematics (NCTM). The NCTM is an organisation of teachers and others concerned with mathematics education, and most of its members come from the United States and Canada, although growing numbers are from other countries. The NCTM currently has almost 100,000 members.

In 1980, the NCTM published *An Agenda for Action,* which made a number of recommendations, most prominently that problem solving be the focus of school mathematics, with basic skills defined as more than computation. The *Agenda* was NCTM's way of providing direction to the field, but it was also its first major effort to influence public policy. The *Agenda* was both well received and ultimately rather influential in national education policy. The organisation began to realise that it had an important political role to play. Then in 1984, the NCTM Board of Directors appointed a task force to plan the development of comprehensive guidelines for the kindergarten to Grade 12 school mathematics programme. The first project was the writing of *Curriculum and Evaluation Standards for School Mathematics,* published in 1989. Standards for teaching mathematics were dealt with in a later project, leading to *Professional Standards for Teaching Mathematics* in 1991. Four years after that, *Assessment Standards for School Mathematics* was published. These documents, particularly the first, helped launch what became known as the standards movement. The documents attempted to provide a vision of mathematical literacy for today's world.

Through the first half of the 1990s, the reaction to the standards was almost entirely positive. Publishers began to call their textbooks "standards-based". Curriculum development projects, first at the middle school level and then at the high school level, were funded by the U.S. National Science Foundation to develop new instructional materials. States and local districts began to "align" their mathematics curricula with the NCTM standards. The few discussions of the mathematics standards in the media did not take issue with what they said.

THE MATH WARS

Gradually, however, a backlash began to form. The NCTM was charged with promoting a movement labelled *whole math,* a term chosen so that it could be lumped together with *whole language* methods of teaching reading and both then characterised as efforts to subvert education. Another label for the standards-based reform was the *new-new math,* which indicated that it was somehow a successor to the new math of the 1960s, an approach also seen as discredited. Groups of parents and mathematicians were formed to work against standard-based changes in school mathematics. Of these groups, the oldest and most visible is Mathematically Correct, which runs an up-to-date and informative Web site <http://ourworld. compuserve.com/homepages/mathman/index.htm> that presents the anti-reform position in detail. (For details of the movement in California, the backlash there, and its consequences, see Wilson, 2003).

Critics began to use terms like *fuzzy math* to characterise almost anything done in the name of standards-based change. Defenders of that change then accused critics of wanting to return to a *parrot math* curriculum. The media related horror stories of children wasting their time in misguided "explorations" and not learning basic facts. In January 1998, Richard Riley, the U.S. Secretary of Education, weighed into the controversy, calling for a cease fire in what he termed the "math wars" – reviving a term that had been used almost 40 years before to characterise the new math reform efforts (DeMott, 1962: ch. 9).

In response, and to take advantage of the first 10 years of experience with and reactions to the standards, NCTM published in 2000 its *Principles and Standards for School Mathematics.* That document brought together and updated standards for the mathematics classroom, combining curriculum, teaching, and assessment. Because mathematicians had been at the forefront of many of the criticisms of its earlier standards documents, the NCTM included more mathematicians on the writing groups for the 2000 document and made stronger efforts to get prepublication reviews by mathematicians. Much criticism, however, continued to be heard, indicating that there were still some topics on which perspectives diverged.

DIVERGENT COMMON SENSE

Definitions

One topic on which the common sense of mathematicians appears to differ from that of teachers is the role of definitions in teaching mathematics. Mathematicians tend to regard the presentation of a precise definition as essential when students are learning a new concept (Milgram, 2005: 85–92). The definition should be "mathematically accurate" (Mathematics Standards Study Group [MSSG], 2005). For example, in introducing the topic of function, the set-theoretic Dirichlet-Bourbaki definition might be provided so that learners would have an unambiguous criterion to apply. In contrast, schoolteachers tend to be wary of providing a formal definition before learners have some idea of what is being defined (and for which they have developed a personal definition). In the common sense of teachers, the learner needs an image of the concept (Tall & Vinner, 1981; Vinner, 1983) before a formal definition can be understood. Furthermore, teachers are often content to work with a provisional definition until learners have explored the concept and have become familiar with various examples and non-examples.

Algorithms

So that learners can perform computations, they need to learn algorithms for the operations. Algorithms differ greatly in transparency, efficiency, generality, and precision (Kilpatrick, Swafford, & Findell, 2001: 103), and learners need to understand how to find an appropriate balance among these characteristics. A teacher may encourage learners to work with a transparent but relatively inefficient algorithm, for example, so that they can see how the algorithm operates. The teacher may even allow learners to construct and use their own algorithms before moving on to one that is more precise and general. In contrast, mathematicians typically see little value in learning any algorithm that is not what they consider to be a "standard" paper-and-pencil algorithm. They argue against the adoption of "student developed algorithms" (Milgram, 2005: 178). It seems only common sense to learn one algorithm that works efficiently every time rather than using a half-baked procedure. For teachers, however, it may not be so clear which algorithm ought to be taken as standard, particularly when the learners come from families in which the adults went to school in different systems and were taught different algorithms. The teachers' common sense leads them to make sure that learners know and can use "an algorithm that is general and reasonably efficient" (Kilpatrick, Swafford, & Findell, 2001: 414) but not necessarily the same algorithm that a mathematician might label *standard*.

Technology

Technology, and in particular the use of calculators, is a topic on which the common sense within the community of mathematicians and the community of teachers is almost as mixed as it is between the two groups. One line of common sense, promoted by many but far from all teachers, argues for the extensive use of technology in school mathematics because of the power it provides learners to visualise and explore complex situations dynamically. Another line of common sense, however, primarily from mathematicians but also from teachers, argues that learners become "too dependent" on technology when it is used too extensively in school. Learners reach for technology when they should be using their memory and their reasoning abilities.

Calculators appear to pose a particularly vexing issue. A recent statement from a group of mathematicians who were addressing the issue of giving elementary school students practice in performing multi-digit arithmetic operations, said,

> The only role for calculators in this process is to check answers computed by hand. (MSSG, 2005: 4).

The 1989 NCTM standards document had seemed to be promoting calculator use when it said,

> Appropriate calculators should be available to all students at all times. (NCTM, 1989: 8)

But the May 2005 NCTM position statement entitled *Computation, Calculators, and Common Sense* appeared to take a more middle-of-the-road stance, saying,

> The teacher should help students learn when to use a calculator and when not to, when to use pencil and paper, and when to do something in their heads. (NCTM, 2005)

Calculator use appears to be a topic on which common sense has yet to congeal across and within communities.

Statistics
Many teachers in North America have embraced the inclusion of data analysis and statistics in the school curriculum because it allows them to engage students in interesting class activities, including the construction of mathematical models for the data. They find that students enjoy using mathematics to investigate realistic situations involving statistics (Kilpatrick, Hancock, Mewborn, & Stallings, 1996). The 2000 NCTM standards document proposes data analysis and probability as a standard throughout the grades from pre-kindergarten to Grade 12. Some mathematicians, however, take a more jaundiced view of the introduction of statistics into the school curriculum. They worry about schoolchildren wasting their time making histograms of data they have gathered when they could be learning arithmetic. They see statistics in school mathematics as involving little serious work. At a March 2006 conference on issues in school mathematics, a prominent mathematician told the teachers present that they should leave the teaching of statistics to the colleges and universities "where we know how to do it right." Their common sense tells teachers that learners respond well to statistics as part of their mathematics education. Their common sense tells mathematicians that statistics is difficult to teach well and may take up space in the school curriculum better left to important mathematics.

Mathematics for All

One of the most curious themes on which mathematicians appear to diverge in their commonsense views from the teachers in NCTM concerns the latter's efforts to promote mathematics for all. The NCTM 2000 standards document says,

> Equity ... demands that reasonable and appropriate accommodations be made as needed to promote access and attainment for all students. (NCTM, 2000: 12)

The slogan on the NCTM Web site (http://www.nctm.org) proclaims "More and Better Mathematics for All Students." Nonetheless, at least some mathematicians question that idea. Martin Gardner (1998), reviewing a commercial algebra textbook, an NCTM yearbook on equity, and a public television videotape series on mathematics, argues against the NCTM's effort to change school mathematics, calling it both "fuzzy math" and "the new new math." He complains that it

> is heavily influenced by multiculturalism, environmentalism, and feminism. (Gardner, 1998: 9)

Although he rightly complains about the absence of recreational mathematics in the materials under review as well as their over-reliance on cultural artefacts of questionable value or validity, he seems almost to resent the notion that school mathematics might be made more accessible to more students. Such critics of the NCTM's efforts treat school mathematics as akin to a finite resource to be distributed only to the deserving, those willing to work hard to attain it.

SEEKING COMMON GROUND

In December 2004, Richard Schaar, a mathematician and senior vice president of Texas Instruments, invited two mathematicians, James Milgram of Stanford and Wilfried Schmid of Harvard, and three mathematics educators, Deborah Loewenberg Ball of the University of Michigan, Joan Ferrini-Mundy of Michigan State University, and me, to a so-called peace summit in Washington, DC, where we attempted to find common ground in the math wars. After a second meeting at the offices of the Mathematical Association of America (MAA) in June 2005, with numerous e-mail exchanges in between, the group posted a statement on the MAA Web site, and it was published in the *AMS Notices* (Ball, Ferrini-Mundy, Kilpatrick, Milgram, Schmid, & Schaar, 2005). The purpose of the statement was to demonstrate that there was agreement across groups on some of the points being debated in the math wars.

A response by Tony Ralston (2006) called the effort "a valuable exercise" but also concluded that the statement was unexceptional, bland, and ambiguous. It did not, he said, address major points of contention, including curriculum and technology. He argued that before any nontrivial consensus on issues of school mathematics could be achieved, there first needed to be "a level of respect in both communities for the other that will mean that inevitable disagreements need not erupt into shouting matches."

To continue the search for common ground, a meeting of approximately 75 mathematicians, teachers, and other mathematics educators was held at Indiana University-Purdue University Indianapolis on March 2–5, 2006. The purpose was to explore issues in school mathematics on which common ground might be achieved, spreading the conversation beyond the small group that had begun it. Five groups were formed to address the topics of algebra, algorithms, probability and statistics, technology, and teacher preparation. Reports from the groups were posted on the MAA Common Ground Web page (http://www.maa.org/common-ground). Those reports, although far from polished or definitive, indicated that there was, indeed, considerable agreement in the various communities concerned with the teaching of school mathematics.

DEVELOPING COMMON SENSE

The Common Ground report and its aftermath, like that of the National Research Council's Mathematics Learning Study (Kilpatrick, Swafford, & Findell, 2001) or the report of the RAND Mathematics Study Panel (2003), demonstrate that what appear to be divergent commonsense views of various groups concerned with school mathematics can be reconciled, given the right circumstances. Members of different groups can come together and educate their collective common sense.

What does it take? For one thing, it takes time. The 4-page Common Ground statement took 7 months. The report of the Mathematics Learning Study took 2 years. The RAND report took almost 3 years. People need an opportunity to get acquainted with each other's views and understand how they are thinking about the issues. As Ralston's (2006) critique suggested, they need to develop a climate of mutual respect and honest exchange of opinion before they can move forward.

An important feature of these efforts appears to be the engagement by all participants in the production of a written report. That task requires that people listen to one another carefully, ask for clarification or examples when a point is not clear, and formulate language that all can agree on. Working collectively in this way gets people away from the more extreme language that they might use when giving a talk or posting a message on the Internet. It allows them to discover that the issues on which they disagree strongly are likely to be fewer than they had thought and that with sufficient discussion, they can reach some consensus.

Developing a collective common sense across the several communities concerned with the teaching of school mathematics will never be an easy matter. It will always require leadership, support, time, cooperation, communication, and good will. Nonetheless, it appears to be not only a worthwhile goal but also an attainable one.

JEREMY KILPATRICK

REFERENCES

Ball, D. L., Ferrini-Mundy, J., Kilpatrick, J., Milgram, R. J., Schmid, W., & Schaar, R. (2005). Reaching for common ground in K-12 mathematics education. *Notices of the American Mathematical Society, 52*, 1055–1058. Also available and retrieved September 5, 2006, from Mathematical Association of America Web site: http://www.maa.org/common-ground/cg-report2005.pdf

DeMott, B. (1962). *Hells and benefits: A report on American minds, matters, and possibilities.* New York: Basic Books.

Fischbein, E. (1987). *Intuition in science and mathematics: An educational approach.* Dordrecht: D. Reidel.

Freudenthal, H. (1991). *Revisiting mathematics education: The China lectures.* Dordrecht: Kluwer.

Gardner, M. (1998, September 24). The new new math. *New York Review of Books, 45*, 9–12.

Keitel, C., & Kilpatrick, J. (2005). Mathematics education and common sense. In J. Kilpatrick, C. Hoyles, & O. Skovsmose (Eds.), *Meaning in mathematics education.* Dordrecht: Kluwer, 105-128.

Kilpatrick, J., Hancock, L., Mewborn, D. L., & Stallings, L. (1996). Teaching and learning cross-country mathematics: A story of innovation in precalculus. In S. A. Raizen & E. D. Britton (Eds.), *Bold ventures, vol. 3: Case studies of U.S. innovations in mathematics education.* Dordrecht: Kluwer, 133-243.

Kilpatrick, J., Swafford, J., & Findell, B. (Eds.) (2001). *Adding it up: Helping children learn mathematics.* Washington: National Academy Press.

Mathematics Standards Study Group (2005, September). *What is important in school mathematics.* Retrieved September 5, 2006, from Mathematical Association of America Web site: http://www.maa.org/pmet/resources/MSSG_important.html

Milgram, J. (2005). *The mathematics pre-service teachers need to know.* Retrieved September 5, 2006, from Stanford University Department of Mathematics Web site: http://math.stanford.edu/ftp/milgram/FIE-book.pdf

NCTM (1980). *An agenda for action.* Reston: National Council of Teachers of Mathematics.

NCTM (1989). *Curriculum and evaluation standards for school mathematics.* Reston: National Council of Teachers of Mathematics.

NCTM (1991). *Professional standards for teaching mathematics.* Reston: National Council of Teachers of Mathematics.

NCTM (1995). *Assessment standards for school mathematics.* Reston: National Council of Teachers of Mathematics.

NCTM (2000). *Principles and standards for school mathematics.* Reston: National Council of Teachers of Mathematics.

NCTM (2005, May). *Calculators, computation, and common sense: A position of the National Council of Teachers of Mathematics.* Reston: National Council of Teachers of Mathematics. Retrieved September 5, 2006, from the NCTM Web site: http://www.nctm.org/about/pdfs/position/computation.pdf

Ralston, A. (2006, January). K–12 mathematics education: How much common ground is there? *Focus [The Newsletter of the American Mathematical Association], 26*(1), 14-15. Retrieved September 5, 2006, from Mathematical Association of America Web site: http://www.maa.org/common-ground/Ralston-FOCUS-Jan06.html

RAND Mathematics Study Panel (2003). *Mathematical proficiency for all students: Toward a strategic research and development program in mathematics education.* Santa Monica: RAND.

Tall, D., & Vinner, S. (1981). Concept image and concept definition in mathematics with particular reference to limits and continuity. *Educational Studies in Mathematics, 22*, 125–147.

Vinner, S. (1983). Concept definition, concept image and the notion of function. *International Journal of Mathematics Education in Science and Technology, 14*, 293–305.

Wilson, S. (2003). *California dreaming: Reforming mathematics education.* New Haven: Yale University Press.

AFFILIATIONS

Jeremy Kilpatrick
Department of Education,
University of Georgia

BILL ATWEH

INTERNATIONAL INTERACTIONS IN MATHEMATICS EDUCATION

Pragmatics, Problematics and Potentials

International contacts in mathematics have a very long history that preceded the era of globalisation. The transmission of mathematical knowledge from the East (e.g. India) and the South (e.g. Arabia) formed the roots of mathematics as a discipline in Europe (Powell & Frankenstein, 1997). Similarly, in mathematics education, the establishment of the International Commission of Mathematics Instruction (ICMI) in 1908 was both a reflection of the belief that mathematics educational problems can, and need to be solved globally, and at the same time provided promotion of that conviction. With ease of travel and communication and greater awareness of developments and the needs of various countries, contact between mathematics educators has escalated and taken diverse forms. While mathematics educators have always shown an acute awareness of the international status of their profession (as reflected in numerous publications and conferences with the term "international" in their titles), there has been little problematisation of this phenomenon and research activity about the benefits or problems that might arise.

Views expressed by mathematics educators about international contacts and activities vary. For some, international interactions lead to greater awareness and understanding of difference that, leads to assisting the less able, to tolerance and conflict resolution. Often these educators achieve greater conscious understanding of their own assumptions and salient aspects of their own practices. To others, such contacts may lead to homogenisation, colonisation and to the marginalisation of the 'have nots'.

In this chapter, I will build upon my previous work with different colleagues on globalisation and internationalisation in mathematics education (Atweh & Clarkson, 2001; Atweh, Clarkson, & Nebres, 2003) to discuss issues related to international interactions between mathematics educators in particular. I will present a model that might allow us to differentiate between forms of global contacts and give case studies of two collaborations that might be educative in our reflections on our collective and individual practices. In this discussion, I will rely on conversations with some leading mathematics educators from Latin America (Brazil, Colombia and Mexico) and Asia (South Korea, The Philippines, and Vietnam) that I conducted between 2001 and 2002[54]. However, I will commence this discussion with a summary of some issues raised during a Discussion Group at the most recent International Congress of Mathematics Education (ICME) in

U. Gellert, E. Jablonka (eds.), Mathematisation – Demathematisation: Social, Philosophical and Educational Ramifications, 171–186. © 2007 Sense Publishers. All rights reserved.

Denmark 2004. The topic of international cooperation involving mathematics teachers and educators from different countries presented a wide range of views and issues about international activities.

ISSUES IN INTERNATIONAL COLLABORATION

First, there are different, and at times conflicting motivations behind international collaboration. In a globalised world dominated by economic rationality, many of these international collaborations have their roots in financial benefits to the participants. For example, as many universities around the world are facing a reality of reduced government funding, they are turning to international students and projects as a significant source of income. Similarly, many less industrialised countries that depend on international loans to develop their eduction systems and infrastructure often face additional requirements for specific types of 'reforms' that necessitate contacts with overseas educators and systems. Other international collaborations, are based on more altruistic motivations such as the provision of assistance for countries with limited resources to develop their capacity to build their infrastructure and educational reform. Perhaps such collaborations are based on the premise that mathematics is associated with economic development and prosperity, hence assisting poorer countries through establishing a solid mathematics education system may contribute to the reduction of overall poverty. Further, certain types of collaboration may yield direct benefit to individual academics seeking new research sites and markets for their publications.

Second, international collaborations face many factors that limit participation in them by many educators around the world. Not the least of these limitations is financial. The cost of attending international gatherings, or subscribing to international journals is a prohibiting factor for many international mathematics educators from less industrialised nations. Similarly, educators from non-English speaking countries often feel excluded from many international activities that are in English. The final report of the IMCE discussion group identified further problems arising from language:

> In addition to the dominance of English in many international cooperative activities, the problem of language is also a matter of particular professional jargon used in different national communities to refer to the objects of their practices. Problems of understanding emerge due to differences in the meanings of commonly used terms. For example, the phrase "didactics of mathematics" carries almost opposite meanings for a native English speaker and speakers of other European languages. Further, care must be given not to exclude some participants from having access to that technical language by oversimplifying it. Hence, genuine cooperation must include a process of communication in which, through languages (natural and specialized), the parties involved negotiate their meanings and intentions for action. (Atweh, Boero, Jurdak, Nebres, & Valero, in press)

Third, international collaboration may have serious negative effects on some participating countries. Without due care, collaboration between educators with varying backgrounds, interests and resources may lead to domination of the voice of the more able and marginalisation of the less powerful (Atweh, Boero, Jurdak, Nebres, & Valero, 2004). Further, uncritical collaboration may confuse aid to the less resourced countries with a 'missionary' attitude that leads "to a patronizing relationship, which does not respect and value the diversity of the parties involved. Instead, an attitude of humility and openness to learn from each others should be the basis of international co-operation" (Atweh, Boero, Jurdak, Nebres, & Valero, in press). Lastly, non-critical collaboration can lead to the homogenisation of mathematics education around the world and a loss of local variations and cultural difference. As might be expected, the discussants represented divergent views about this point.

Some have pointed out how in the new world order, reforms in one country are transplanted, in many cases uncritically, to other countries. Some talked about the "Americanisation" or the world curricula. However, many also argued that cooperation between different countries can lead to awareness of different approaches to both research and teaching methods that might increase the variety at local level. Researchers in the discussion pointed out that the mathematics education literature in many countries reflects a greater variety in methodologies and theoretical stances now than fifty years ago. Hence, at the same time that trends in research and teaching are becoming homogenized at a global level, they are becoming increasingly diversified at a local level. (Atweh, Boero, Jurdak, Nebres, & Valero, in press)

MODEL FOR STUDYING INTERNATIONAL INTERACTIONS

In previous publications (Atweh & Ragusa, 2003; Ragusa & Atweh, 2003), based on theorisation by Young (1990) and Fraser (1995), a colleague and I discussed a model for social justice as it relates to international collaboration. This model is used here to discuss different types of international collaboration and to identify some issues they raise.

Young's main critique of traditional conceptions of social justice suggests it is based on "having" rather than "doing." Young argues grounding social justice in individual solutions allows little room for consideration of divergent social groups. Hence extending traditional models based on the *distribution* of material goods to disadvantaged individuals, to other goods such as self-respect, honour and opportunity for disempowered social groups, is problematic. To understand the struggles for social justice by a variety of groups, such as women, African Americans, and gay and lesbian people, feminist theorists created a discourse of social justice based on *recognition*. Fraser (1995: 68) expounds:

Demands for "recognition of difference" fuel struggles of groups mobilised under the banners of nationality, ethnicity, 'race', gender and sexuality. [...]

173

And cultural recognition replaces socioeconomic redistribution as the remedy of social injustice and the goal of political struggle.

Fraser goes on to argue that social justice requires both redistribution and recognition measures. She argues that affirmation and distribution cuts across the redistribution-recognition divide. *Affirmative* remedies include those "aimed at correcting inequitable outcomes of social arrangements without disturbing the underlying framework that generates them" (Fraser, 1995: 82), while *transformative* remedies are "aimed at correcting inequitable outcomes precisely by restructuring the underlying generative framework" (ibid.). Based on this discussion, we put forward a model comprised of four modes characterising possible interactions among academics from different cultures.

Table 1. A model of possible interactions

	Affirmation	*Transformation*
Redistribution	Mode 1: Aid Attributes: Sharing of information and resources among countries. Represents cultural classification based upon access to knowledge. Can generate misrecognition.	Mode 2: Development Attributes: Restructuring of relations of knowledge production. Blurs group identification. Can help remedy misrecognition.
Recognition	Mode 3: Multiculturalism Attributes: Acknowledging cultural differences, such as cross cultural research. Supports group identification.	Mode 4: Critical Collaboration Attributes: Deep restructuring of relations of recognition. Blurs group differentiation.

I will turn now to discuss each type of interactions in some detail.

The aid mode

In the period immediately following World War II, some of the allies were given a mandate to assist with the development of independence and modernisation of less developed countries. Specific examples included large-scale aid programs for the building of infrastructure and curriculum development at the teacher education and school level. Such programs continue today with many 'more affluent' countries providing varied projects to assist 'less affluent' countries in their development and modernisation. For example, in the financial year 2005-2006 the Australian government provided AUS$2.5 billion as assistance (AusAID, undated) to less industrialised countries in the Asian Pacific region.

In the global context of increasing the inter-dependence of countries and economies and in the context of increasing gaps between the 'haves' and 'have nots' such activities are imperative and have contributed to the needs and

174

aspirations of many struggling countries. In a dialogue with educators from Colombia (Atweh, Clarkson, & Nebres, 2003), I noted the sense of despair and isolation that educators felt in the country about the limited resources available to participate in international contacts. The Colombian educators were well aware of their country's richness in human resources for finding solutions to their own problems. Yet, that potential is not reached because of the limited financial resources. Colombia has to support local research. This gives rise to a situation where "what we have done is to consume without assessing what has been produced in schools from other countries". When asked what their expectations were from international contacts, the participants in the focus group were very direct and candid in their reply. They aspired to more internationally financed research or at least to co-financed research projects.

However, international aid programs are based on the model of transmission of goods, (either material or symbolic) from one culture to another. These models cause serious concern. To start with these forms of knowledge transmission often lack reciprocity among the players leading to a form of colonisation of mathematics education from the North to the South and from West to East. In describing different Southeast Asian countries' curriculum and school structures Ben Nebres, a leading mathematics educator in Southeast Asia, noted they reflect the chequered colonial history of the different countries: "The mathematics education curricula and the education systems in many cases were transplants from the colonial countries" (Atweh, Clarkson, & Nebres, 2003). Similar patterns can be observed in research conducted in mathematics education. In a dialogue with mathematics educators from the Philippines, one educator discussed how international contacts determine the type of research that is conducted in their country where "the research staff [being] very much influenced by what they see in [overseas] journals, and sometimes [concentrate their research] rather than on something that will improve [the local conditions, they concentrate on] on trivial topics [for our context]". She explained how some of the more serious problems relate to poverty in the midst of urban society and rural areas – concerns about which traditional research in mathematics education both internationally and locally are silent. Another educator lamented, "I think like in any globalisation, many of us are torn between engaging in these global activities and at the same time trying to preserve whatever Filipino culture we can identify ourselves".

Development mode

Ragusa and Atweh (2003: 3) defined 'development' as a

> transformative process whereby goods and/or knowledges are distributed across social structures, groups and/or individuals. Development seeks to change pre-existing patterns and norms of knowledge production and may have short or long-term effects. However, it does not necessarily problematise differences in interests and needs of the different participants.

International interactions under this model include international postgraduate students in developed countries and programs that contribute to the professional development of educators. Such activities may contribute towards the long-term empowerment of professionals within less developed countries.

In this context, we argue that while development of expertise is a socially just endeavour, the lack of recognition of and respect for difference implies that such programs result in the reproduction of current practice and thinking on a global scale. Many of the doctoral holders from the participating countries, in particular developing nations, obtained their qualifications from overseas countries. Increasingly, more and more countries are developing their own PhD programs. For example in the 1990s, a few academics from Colombia were successful in obtaining scholarships to undertake doctoral programs at overseas universities. In the mid-Nineties, as a result of collaboration between five private and public universities in the capital Bogotá and some regional cities, a national doctoral program in science and mathematics education commenced. Candidates in the program have to demonstrate a mastery of at least one language other than Spanish. This is an attempt to encourage their contribution to international conferences and publications.

Undoubtedly, doctoral students studying abroad bring back to their countries theories and methodologies from their host countries. For example, in Korea, about half of the educators working in mathematics education at universities have obtained their qualifications from the United States. According to one participant, "that's why the Standards affect us so much because we are used to that curriculum and we studied there and we come back and [it is the model we use] whenever we talk about the curriculum. [...] So we are teaching and researching in Korea, but our minds are over there ... because we got all our basic ideas from the States".

Lastly, there is a problem of 'brain drain' from developing nations to developed nations (UNESCO, 1998). The educators from the Philippines talked about hundreds of qualified and experienced teachers leaving for jobs overseas, going in particular to the United States. I recall in the late 1990s, while being involved in a professional development project for staff from institutes of teacher education in The Philippines, meeting many young and aspiring mathematics education. When I returned to the country in 2003, most, had already left for overseas work. Naturally, there is a human cost for the individual and their families when leaving a family orientated country. However there is also a significant financial cost to the country for replacing these teachers. Nevertheless, the overseas offers are very tempting for people who "even with their PhD degree are taking home something like Aus$200 per month".

Multicultural mode

Potentially, all international contacts can be mutually beneficial and contribute to a greater awareness between the participants and of their own practices. Such collaboration may lead to self-examination of practices, assumptions and values and further to the creation of ways for dealing with educational problems.

International interactions based on distribution, such as those discussed in the above two categories often result in the participants being divided into those who help (persons from developed countries) and those who need help (persons from developing countries). In other words, even when collaborative reciprocity is strived for, it is often hard to achieve in practice.

Ragusa and Atweh (2003: 4) defined 'multiculturalism' as a mode that

acknowledges differences among cultures and supports multiple identities. However, it does not seek to alter or change access to, or production of, material and/or symbolic goods.

Perhaps an example of multicultural interactions in mathematics education is the international movement of ethnomathematics (Atweh, Clarkson, & Nebres, 2003). Arguably, international comparative studies on mathematics curriculum are not very common in mathematics education. However, within the past three decades, mathematics education has witnessed an increase in cross-national comparative studies on curriculum and student achievement. These studies have received considerable attention within and outside the field. International testing has been widely covered by media and featured in public debates about education. The potential benefits, and problems, with international testing have been addressed elsewhere (Clarke, 2003; Kaiser, Luna, & Huntley, 1999; Robitaille & Travers, 1992). The discussants in the focus groups identified several social justice issues faced by educators in their countries directly resulting from the international testing of achievement.

An outcome of international testing, and the accompanying media frenzy, was the introduction by many countries of testing based upon educational reforms. A leading educator from Brazil talked about "a testing epidemic" hitting the country. Focus on test results has the potential to give an inaccurate, and damaging, impression about what constitutes mathematics. This is the "perverse" side of a globalisation-based utilitarian understanding of mathematics that serves the interests of big business and global competition. It leads to an uncritical adoption of curricular focus from one culture to another. An educator from Brazil cites the example of the number of school districts in the United States that adopted texts and curricula from Singapore in the belief that they had a huge success when subjected to international tests.

Unproblematic use of international testing does not take into account the context of the educational systems in the particular countries. For example, students from the Philippines participating in these studies are one year behind many of their counterparts around the world because of the starting age of formal schooling. An educator from the Philippines questioned the use of international tests on significant changes to education systems in many developing nations. The participants noted that one tangible benefit of international testing is that the tests challenge teachers and educators to look at their testing practices and compared them to instances where the government provided training for teachers in test development techniques that assess higher order skills. However, considering the reality of Philippines classrooms, where many classes contain up to sixty students, and six students share one text book, the mass hysteria about testing results is "not going to make a big dent in your performance next time unless you tackle the basic problem of the maths [teaching resources]".

Finally, Keitel and Kilpatrick (1999) raise several political questions about such international comparative studies. They argue that the outcomes of these studies are perceived as biased towards the host country; that is, of those who do the data collection, the analysis and the funding. These authors question if this is to the detriment of other countries and their concerns about improving education systems. Outcomes of such studies are also perceived as necessarily reductionist, as results cannot do justice to the very complex factors involved. The authors claim that the mathematical tasks do not represent the curricula taught in many schools, teachers' questionnaires do not represent the whole range of teaching practices, and the results do not offer valid comparisons between the various countries' curricula with their divergent cultural and social contexts. "No allowance is made for different aims, issues, history and contexts across the mathematics curricula of the systems being studied" (Keitel & Kilpatrick, 1999: 243). They conclude that comparative testing is not really useful as an educational tool, as it does not produce a clear view of what's really happening in the classroom and why.

Critical mode

Ragusa and Atweh (2003: 4) observe that critical collaboration

> entails the deep restructuring of social structures and relations. It is a dynamic, dialectical process for assessing the ability to transform and change norms, political systems and codes of practice. Critical collaboration recognises difference and creates a forum for authentic dialogue.

Like multiculturalism, critical collaboration aims to give recognition and respect to the knowledges different cultural groups and countries provide. However, in this category effort is made to challenge the structures that give rise to inequality in status, as well as the knowledge shared, among nations. Critically collaborative activities are necessarily based on participation from educators in different countries as all work to develop local knowledge and simultaneously contribute to collective international knowledge.

In the discussion group at the ICME conference, the participants discussed the difference between genuine and nominal collaboration in international activities. The report states:

> Considerable discussion was focused on the terms equitable and genuine in relation to international co-operations. Many participants warned against the naive position towards the meaning of international cooperation that pretends that cooperation necessarily implies they are carried out among equals. Often, international co-operations are established among unequal participants with some participants positioned in a dominant role due to access to resources such as funds, technology or expertise in dominant modes of operation in

research and/or teaching in mathematics education. In these contexts, equality in cooperation is built on a respect for the different type and not quantity of contributions of the partners, on the acknowledgement of the equal value given to the different knowledges the participants, and on the necessity to tackle problems of relevance for each of the parties involved. Moreover, genuine collaboration is one that is based on self-critical reflection by the different partners about their self interests and expected contribution to the cooperative activity, and on the transparency among participants in relation to their expectations, contributions, benefits and voice in representation of the results. (Atweh, Boero, Jurdak, Nebres, & Valero, in press)

Before I leave this discussion about the different modes of international collaboration, it is useful to point out that the different categories are not to be taken as disjointed and mutually exclusive. Critical collaboration does necessarily involve elements of aid, in particular when such aid allows the less resource rich countries to participate. It also involves professional development, not only for the less affluent countries but to all participants. Likewise it includes respect for differences, but enters into a critique of difference based on the interests of all participants. In the following section, I will discuss two case studies of international collaboration that illustrate some of the issues raised in this section.

CASE STUDIES OF INTERNATIONAL COLLABORATIONS

Case study 1: An international knowledge network[55]

The Learners Perspective Study (LPS) is an informal network of international mathematics educators involved in a collaborative project investigating classroom interactions in mathematics classrooms. The initial idea for the project stemmed from an informal conversation during an international conference between David Clarke, from Australia and Christine Keitel from Germany. The discussion centred around some of the limitations of the Third International Mathematics and Science Study (TIMSS) video study. Among their concerns about the TIMSS data collection methods were its lack of ability to capture student-to-student discussions in the classroom and access students' perception of teacher actions and classroom events. The agreed aim of the LPS project was to develop a means of collecting data from the three countries involved in the original TIMSS video study – Germany, Japan and the United States – plus Australia. Yoshinori Shimizu was recruited from Japan, and Joanne Lobato from the US to allow for validity of data collection from those countries. Initial project funding was obtained from the four participating countries. As discussions developed about the project, the project's scope expanded to include more countries. For example, Sweden expressed an interest in participating and then, through further individual contact and discussion, the project extended to include Hong Kong, Mainland China, Israel and the Philippines.

The participation by the Philippines is particularly useful for our discussion here. Although the Philippines' educators wanted to join the international team, they were concerned about the lack of Filipino funds available to conduct such a study that would allow them to participate at the group's international meetings. However, to facilitate such participation, other project participants elected to subsidise the Philippines by sending them equipment previously used in the Australian data collection. In addition, two technicians were sent to train educators to operate the equipment. Further specialised training in conducting interviews was provided by the Australian team in Manila. Finally, Australian funds were used to subsidise the Philippines' participation at the international research team meeting.

The project data was generally subjected to three types of analysis. First, a project-wide analysis was conducted in accordance with the mutually agreed project aims. This analysis was conducted on project-wide categories, such as lesson structure, and was based on Clarke's earlier work in the 'Negotiation of Meaning' project. Second, at times, the participating countries sub-divided into groups, according to specific interests, and special analysis was performed on their data. For example, Hong Kong and Sweden were interested in theories of variation; Germany and South Africa, focused on social justice; the United States and Sweden explored issues related to mathematics as a discipline; and Australia and Hong Kong were concerned about issues of knowledge generation in the classroom. Finally, individual countries and researchers had the option to perform analysis on their own data.

At the initial meeting of the group, some apprehension was expressed by representatives of poorer countries, that rich countries due to their greater resources, may "appropriate" their data by completing analyses more efficiently. To address this concern, the group developed stringent gate-keeping mechanisms to safeguard each country's data from the others. Data from one country could only be used by another with the permission of the first country's researchers. Intended data users were expected to send a draft of any publication making use of the data to the representative for approval. This ensured the data was not misinterpreted and that it would not have a negative effect on the "owner" country.

While participants in this project had differing levels of previous experience in research publication and access to facilities, the project provided a professional learning experience for all participants. More experienced researchers gained access to wide data sources; however, the project provided new challenges to their views about classroom teaching and learning, as well as research methods and processes. Similarly, less experienced researchers with limited access to resources, gained access to international forums and training in research and publishing. In addition, all involved learned invaluable lessons about the stresses and realities that accompany working in a multi-national and multi-cultural research team.

The groups became aware of cultural sensitivities and at times annoyances, and different methods of communication. These were often dealt with by the groups on a case-by-case basis. In short, team meetings became a venue for significant learning experiences and an ongoing forum bringing sensitisation and awareness of political and cultural issues of significance to each research group and country. The sharing of common interest and the dedication of the team to working with each other assisted in dealing with the cross-cultural issues. However, good will was not sufficient; deliberate and open self-critical discussions during the international meetings proved to be highly valuable in developing a real sense of community between the participants.

Reflection on the LPS project using the above model. This case study illustrates several issues that may arise during collaborations among academics with varied interests, backgrounds and cultures, as well as experiences in research and access to resources. In order for this global collaboration project to include less affluent cultures, sharing of financial burdens was a prerequisite to collaboration. Hence, part of the project can be classified within the aid mode. However, the project also contained elements of the development mode for researchers from less experienced countries. Arguably, the contributions made by different researchers were not equal because the initial model for gathering and analysing the data was driven by the more affluent countries. However, experienced researchers from more affluent countries also experienced professional development as a result of mentoring developing countries. They gained knowledge and appreciation of different research and mathematics teaching traditions. Such collaborations reflect the multicultural mode. Finally, also it is argued that the project revealed certain elements of a critical mode in its dealing with safeguards against possible data "appropriation" by the richer countries. However, through the critical mode lens on this project, one can argue that the research questions posed and procedures followed represent interests that originated in the more affluent countries.

Case study 2: Person-to-person long term collaboration[56]

This is a story of collaboration between two academics from two different continents: Marcelo Borba from Brazil and Ole Skovsmose from Denmark[57].

The two mathematics educators first met in 1992 at the International Congress of Mathematics Education conference in Quebec. This was not an accidental meeting. Previously, Marcelo had read a draft chapter by Ole on "reflexive knowledge" or "technology knowledge" and had written to Ole to discuss some aspects of the paper. Through their correspondence, they found some mutual areas of interest and since both were going to ICME, they decided to meet at the conference. Unknown to each other they shared descriptions of each other to facilitate identification of each other. In the midst of a very busy schedule at the conference, they still found the time to share several informal chats. Seeking further collaboration, Marcelo invited Ole to visit Brazil. Through funds from Danish government, Ole made his first trip to the country in 1994. Marcelo recalls that these plans were made before email was readily available at universities in Brazil. There was a single email address for the whole university which was accessible at ten-minute walk from his office. To maximize the benefit from the visit, plans included a seminar on research and developing research projects for post-graduate students at the university. This collaboration resulted in the publication of the article "The Ideology of Certainty in Mathematics Education" in the journal *For the Learning of Mathematics* (Borba & Skovsmose, 1997).

This working relationship continued for several years and included several visits by Ole to Brazil, visits by Marcelo to Denmark and the delivery of post graduate seminars at Danish universities with further collaborative writing. Naturally, the increased access to email and frequent international conferences facilitated this collaboration.

Elements for successful collaboration. This collaboration did not stem from formal agreements between countries or universities. It was not a strategic collaboration based on existing international projects. It was grassroots collaboration between two academics that shared common interests and willingness to work together. Marcelo recalled several factors that made this collaboration successful and long lasting.

First, both academics came from different research traditions and philosophical perspectives. At the start of the collaboration, Marcelo had expertise in ethnomathematics, a strong movement in Brazil, while Ole was working on critical mathematics, a strong movement in Danish educational discourse. Likewise, Marcelo came from radical constructivist schools of thought while Ole was influenced by the Frankfurt school of critical theory. However, in spite of their differences they shared a commitment to progressive education and a desire to critically study the different traditions as they contribute to political and social dimension of mathematics education. They were both interested in the implications of some of Freire's writings about mathematics education.

Undoubtedly, the different intellectual backgrounds of each academic gave rise to some conflict in their writing. Marcelo recalls the lengthy debate about his accustomed notion of reality stemming from his constructivist background and Ole's adherence to critical theory. At times there were divergent views about tactics in being critical on the writings of other researchers. These differences often led to lengthy debates, a divergence of views and at times compromises resulting in leaving some material out of their collaborative writing. Marcelo identifies humour and respect to each other differences as key requirements for continual collaborations.

Secondly, there were tangible benefits flowing to each collaborator's agenda. Marcelo pointed out the benefit he gained in the area of international exposure. Through this collaboration he was able to publish in several leading international publications. Likewise, Ole had his works translated to a foreign language. Likewise, each partner was a critical friend for their drafts publications. Undoubtedly each partner benefited from expanding their knowledge about other theories informing mathematics education. Although the benefit is mutual when collaborating internationally there may be imbalance. In this case, while it was possible for Ole's work to be translated into Brazilian, some of Marcelo's books for teachers still have to find a Danish publisher. Of course the reasons for this are socio-political and fell far beyond the means of either collaborator.

Overcoming some hindrances. Due to many factors genuine international collaborations are often difficult. This particular collaboration illustrates how some of these challenges may be overcome or negotiated.

First, as argued above, genuine collaboration between partners, be it between countries or individuals, with different access to resources may be problematic. Often available resources limit many academics from developing countries in participating in international conferences and projects. As in *Case study 1* above, this collaboration was made possible by using funds from more affluent society to support the collaboration. More affluent countries often have research and professional development funds that facilitate academic exchanges. However, unequal financial contribution of the different partners to the collaboration, in turn, gives rise to another serious challenge that relates to unequal power relationships that might develop as the collaboration progresses. As this case study demonstrates, negotiating the terms of the collaboration at early stages can be a safeguard against the colonisation by the "haves" of the "have nots". Similarly, maintaining a critical gaze on the development of the collaboration is essential to maintain equal power relationships. The more the parties are aware of their mutual contributions and the benefits that they are accumulating from the collaboration, it is less likely that the unequal contribution of resources would lead to unequal power relation[58].

Second, genuine collaboration is often limited by language differences. This particular case is interesting because the language of collaboration was the second language for both partners. Undoubtedly both academics mastered language to highly functional level. However, problems with the collaboration might have arisen when the academics visited each other's universities to give seminars. This might have caused more concerns in Brazil than Denmark where most post graduate students had a good command of English. The position of English as means of communication in international collaboration is a contested area. We will address it in more detail below. However, using translators during the collaboration went a long way towards overcoming that particular challenge. As Marcelo points out, several different translators were recruited for Ole's presentations in Brazil.

Finally, Internet and email has facilitated communication for international collaboration. The partners in this collaboration have co-authored most of their joint publications through email. Undoubtedly their face-to-face meetings at international conferences and during their visits to each other's countries have provided opportunities for much in depth discussion about their differences. More importantly, face-to-face meetings allow for the development of personal relationships that are essential for long-term collaboration. However, with the ever increasing power of communication technologies, an increasing range of communication are made possible, including video conferencing that simulate face to face meetings.

CONCLUDING REMARKS

In conclusion, I will revisit and discuss the three Ps in the title of this chapter. At a *pragmatics* level, it is safe to say that international contacts will continue to increase in frequency and magnitude for a long time to come. Mathematics education is both a reflection of more general globalisation in our times and a major contributor to it (as reflected in the current debates about international standards and numeracy in many countries). Increasingly, mathematics education is seen as a high priority of many governments aiming to develop competitive economies and for attaining higher standards of living (Kuku, 1995). For some, it serves the neo-liberal agendas of the new world order. However, for others it serves as a powerful tool for empowerment and democratisation of a world marked by inequality and injustice (Skovsmose & Valero, 2001). All agree about the need for increasing international collaboration, albeit of different types, to achieve their agendas.

Different, and often conflicting, agendas are often at play in international collaborations. Some countries and universities are increasingly dependent on economies based on the export of knowledge. Academics are under pressure to play the game and contribute to them. However, some make use of these opportunities to promote other more empowering agendas. The challenge for professional mathematics education is to critically examine personal agendas behind engagement when pursuing international contacts and to reflect on their own practices, and the outcome of these practices in the short and long term. In the midst of pragmatics, there is an opportunity for critical reflection.

At a problematics level, the magnitude and scope of international collaboration has the potential to affect the practices and outcomes for mathematics education internationally. Not all countries benefit equally from such international collaborations. Due to limited resources many academics from less industrialised countries are excluded from participation. Many are not contributing to international debates from their own experience because they are marginalised. Further solutions that may work in one context may not be transplanted uncritically into another. When resources are limited in less affluent countries, this might lead into colonisation. While aid projects are not sufficient by themselves to assure the development of independence of recipient countries, and might have negative consequences, they may be pre-requisites for more meaningful participation.

Secondly, the role of language in international collaborations is quite problematic. Undoubtedly, English has become an international standard. However, it is not an international language. The vast majority of academics in mathematics education do not speak English, although there are few from every country that do. These should not be the only academics that are able to participate in international collaboration. International conferences should aspire to provide translation facilities into other languages. Further, presentations at conferences in more than one language, as available in some European conferences, should become the standard. Hopefully, that will extend to publications as well. I look to a world where more mathematics educators become bilingual. It is worth noting that many educators from non-English speaking countries are, at least, bilingual. Very few from English speaking countries are!

Finally, the potentialities, for international collaboration are a great opportunity to develop a more socially just world. In spite of the potential hijacking of the discipline by economic and global capitalism, our passion and belief that mathematics education is worthwhile both for personal and national empowerment should not subside. Through critical global collaboration as discussed above, this contribution of mathematics to the solution of world problems is more achievable. If mathematics is not part of the solution then it is part of the problem.

Globalisation of issues in mathematics education may lead to a silencing of local concerns and the disappearance of local knowledge. Here, I do not take the stance that local knowledge is necessary good just because it is local. In agreement with Vithal and Skovsmose (1997) what is needed is a critical engagement with local knowledge.

NOTES

[54] The interviews were part of a project was supported by a grant from the Australian Research Council conducted in conjunction with Philip Clarkson.

[55] Data in this section is based on an interview with David Clarke about the LPS project and was triangulated with a discussion with Christine Keitel.

[56] Data in this section is based on an interview with Marcelo Borba about his collaboration with Ole and was triangulated through email discussion with the latter.

[57] In order to highlight the human dimension of this collaboration, I will diverge from the normal academic convention in this section by referring to academics by their first name.

[58] Marcelo recalls that at the beginning of the collaboration he jokingly teased Ole "I want no Vikings here dominating the Amazon again. ... We need to have genuine collaboration".

REFERENCES

Atweh, B., Boero, P., Jurdak, M., Nebres, B., & Valero, P. (2004). International cooperation in mathematics education: A discussion paper. The Tenth International Congress of Mathematics Education (ICME10), Discussion Group 5, Denmark.

Atweh, B., Boero, P., Jurdak, M., Nebres, B., & Valero, P. (in press). International Cooperation in Mathematics Education: Promises and Challenges. A report on The Tenth International Congress of Mathematics Education (ICME10) Discussion Group 5, Denmark.

Atweh, B., & Clarkson, P. (2001). Internationalisation and globalisation of mathematics education: Towards an agenda for research/action. In B. Atweh, H. Forgasz & B. Nebres (Eds.), Sociocultural research on mathematics education: An international perspective. Mahwah: Erlbaum, 77-94.

Atweh, B. Clarkson, P., & Nebres, B. (2003). Mathematics education in international and global context. In A. J. Bishop, M. A. Clements, C. Keitel, J. Kilpartick & F. K. S. Leung (Eds.), Second international handbook of mathematics education. Dordrecht: Kluwer, 185-229.

Atweh, B., & Ragusa, A. (2003). Issues in Social Justice in International Collaborations: Views of educators from around the world. Conference paper presented at The Australian Sociological Association annual conference, Armidale, University of New England, November 2003.

Australian Government AusAID program (undated) (http://www.ausaid.gov.au/) (14/2/2006).

Borba, M., & Skovsmose, O. (1997). The ideology of certainty in mathematics education. For the Learning of Mathematics, 17(3), 17-23.

Clarke, D. J. (2003). International comparative studies in mathematics education. In A. J. Bishop, M. A. Clements, C. Keitel, J. Kilpatrick & F. K. S. Leung (Eds.), Second international handbook of mathematics education. Dordrecht: Kluwer, 145-186.

Fraser, N. (1995). From redistribution to recognition: Dilemmas of justice in a postsocialist society. New Left Review, July-August, 68-93.

Kaiser, G., Luna, E., & Huntley, I. (Eds.) (1999). International comparisons in mathematics education. London: Falmer.

Keitel, C., & Kilpatrick, J. (1999). The rationality and irrationality of international comparative studies. In G. Kaiser, E. Luna & I. Huntley (Eds.), *International comparisons in mathematics education.* London: Falmer, 241-256.

Kuku, A. (1995). Mathematics education in Africa in relation to other countries. In R. Hunting, G. Fitzsimons, P. Clarkson & A. J. Bishop (Eds.), *Regional collaboration in mathematics education.* Melbourne: Monash University, 403-423.

Powell, A., & Frankenstein, M. (Eds.) (1997). *Ethnomathematics: Challenging eurocentrism in mathematics education.* Albany: SUNY Press.

Ragusa, A., & Atweh, B. (2003). Analysing global collaborations using a post-structuralist model of social justice and social change. Paper presented at the Social Change in the 21st Century conference, Carseldine, QUT, Centre for Social Change Research.

Robitaille, D. F., & Travers, K. J. (1992). International studies of achievement in mathematics. In D. Grouws (Ed.), *Handbook of research on mathematics education.* New York: Macmillan, 687-709.

Skovsmose, O., & Valero, P. (2001). Breaking political neutrality. The critical engagement of mathematics education with democracy. In B. Atweh, H. Forgasz & B. Nebres (Eds.), *Sociocultural research on mathematics education: An international perspective.* Mahwah: Erlbaum, 37-56.

UNESCO (1998). World declaration on higher education for the twenty-first century: Vision and action. [www.unesco.org/education/educprog/wche/ index.html]. (5/5/2001)

Vithal, R., & Skovsmose, O. (1997). The end of innocence: A critique of "ethnomathematics". *Educational Studies in Mathematics, 34,* 131-157.

Young, I. M. (1990). *Justice and the politics of difference.* New Jersey: Princeton University.

AFFILIATIONS

Bill Atweh
Queensland University of Technology
Brisbane

PHILIP J. DAVIS

THE MEDIA AND MATHEMATICS
LOOK AT EACH OTHER

INTRODUCTION

Whatever has been learned about how to get at the curve of someone else's experience and convey at least something of it to those whose own [experience] bends quite differently, has not led to much in the way of bringing into intersubjective connection [these two people.]

(Anthropologist Clifford Geertz)

Mathematicians complain that media treatments of their subject are scanty. The media complain that mathematicians give them nothing that is comprehensible or arouses interest. My complaint is that media treatments (from papers to novels to movies and plays) go for the sensational and avoid the mathematisations that characterise contemporary life. In any case, to what extent do the media treatments affect the public's conception of mathematics?

In this article, the word *media* will mean not only the newspapers and television, but also novels, stories, plays, movies, museums; in short, all modes of popular communication of that might shed light on mathematics.

FIRST: A FEW SELECTED A FEW SELECTED VIEWS ABOUT MATHEMATICIANS

(1) Jonathan Weiner, a Pulitzer Prize winning science writer, sent me a personal anecdote. Some years ago, he and his wife took their two boys out for a meal at one of those family restaurants where the tables are covered with paper tablecloths and there are crayons for the kids. While they waited for their orders, the boys entertained themselves by doing mathematics together – scribbling equations on the tablecloth. This impressed the waitress. She stared at the boys slack-jawed. Then she asked the older boy what he wanted to be when he grew up. "A writer," he said. The waitress's jaw dropped even lower. "You're so smart," she said, "and you want to be a writer?"

(2) Go back three centuries and ask what Jonathan Swift's view of mathematics was. In *Gulliver's Travels* (1726) he mocks the Laputian mathematicians:

And although they are dexterous enough upon a piece of paper in the management of the rule, the pencil, and the divider, yet in the common actions and behaviour of life, I have not seen a more clumsy, awkward, and

U. Gellert, E. Jablonka (eds.), Mathematisation – Demathematisation: Social, Philosophical and Educational Ramifications, 187–196. © *2007 Sense Publishers. All rights reserved.*

unhandy people, nor so slow and perplexed in their conceptions upon all other subjects, except those of mathematics and music. They are very bad reasoners, and vehemently given to opposition, unless when they happen to be of the right opinion, which is seldom their case. Imagination, fancy, and invention, they are wholly strangers to, nor have any words in their language by which those ideas can be expressed; the whole compass of their thoughts and mind being shut up within the two aforementioned sciences.

(3) Fast forward to Arthur Conan Doyle's *The Final Problem* (1893) where the mathematical genius is a criminal:

[The arch-criminal Col. James Moriarty] is a man of good birth and excellent education, endowed by nature with a phenomenal mathematical faculty. At the age of twenty-one he wrote a treatise upon the binomial theorem, which has had a European vogue. On the strength of it he won the mathematical chair at one of our smaller universities, and had, to all appearances, a most brilliant career before him.

A brilliant career in the mid 1800's on the strength of the binomial theorem? Does this indicate ignorance, naiveté or satire on the part of Conan Doyle? One might rescue Doyle from ignorance by saying that Moriarty, in advance of his time, worked on the summability of divergent binomial expansions.

THE COMPLAINTS OF MATHEMATICIANS

Briefly: mathematics gets very little coverage of recent developments in the papers. It would like much more. And when mathematics is treated by the papers, they frequently get it wrong. Technical details are mostly omitted and their "take" or point of view is irksome. Years ago, Ronald Rivlin, an applied mathematician and a world authority on rheology (the study of the deformation and the flow of matter) was invited by the *Scientific American* to write an article on his specialty. He did so, but his article was edited in a way that displeased him. A controversy with the magazine editor ensued and the upshot was that Rivlin asked his name to be taken off the article.

Dr. James Crowley, CEO of SIAM (Society for Industrial and Applied Mathematics) wrote me:

Media professionals (writers, reporters) will often tell you that what they seek are items that have "human interest" or that deal with controversial issues. Unfortunately, for mathematics, "human interest" usually involves a stereotypically nerdy or flaky representation of a mathematician. And controversy [which is stock in trade of the papers] is hard to find within mathematics. We tend not to make statements of global warming or stem cell research, but rather whether a given theorem is true or not – hardly the kind of thing to generate heated debates among people on the street.

My main complaint about the treatment of mathematics in the media is that mathematics is not treated in the media. One need only look at the

Science Times on Tuesdays to verify that there is little covered in our area of research. I think this is a general problem, though, and not entirely the fault of the media itself. Look at the press releases issued by the National Science Foundation. One sees very little having to do directly with mathematics or mathematicians.

In 1994, Fields Medallist William Thurston wrote a well received article in the *Bulletin of the American Mathematical Society.* Here are two quotes:

We mathematicians need to put far greater effort into communicating mathematical ideas. To accomplish this, we need to pay more attention to communicating not just our definitions, theorems, and proofs, but also our ways of thinking. We need to appreciate the value of different ways of thinking about the same mathematical structure.

We need to focus far more energy on understanding and explaining the basic mental infrastructure of mathematics. [...] This entails developing mathematical language that is effective for the radical purpose of conveying ideas to people who don't already know them.

Displeasure with the media and its effect on public understanding extends also to scientific reportage. Harvard Professor of Physics Lisa Randall writes:

It would be better if scientists were more open about the mathematical significance of their results and if the public didn't treat mathematics as quite so scary. [...] A better understanding of the mathematical significance of results and less insistence on a simple story would help to clarify many scientific discussions.

Easier said than done. A review in the non-technical *New Yorker* magazine of Randall's newly appeared *Warped passages: Unraveling the mysteries of the universe's hidden dimensions* implied that the details of the physics weren't getting through and the mathematics was relegated to an appendix.

A reviewer in the *Scientific American* of *The Best American Science Writing,* 2004 writes:

Biology, physics, biotechnology, and astronomy, to anthropology, genetics, evolutionary theory, and cognition, represent the full spectrum of scientific writing from America's most prominent science authors, proving once again that 'good science writing is evidently plentiful'.

Where is mathematics in this list of the best science writing?

THE MEDIA'S POINT OF VIEW

One day a local ABC-6 TV reporter came around to the offices of the American Mathematical Society (AMS) in Providence, R.I.. He looked at the display in the lobby and then spoke to a staff member of the Society who does "public awareness." He said to her, "I've passed this building many times and I've often wondered what on earth goes on inside. But I really don't want to know." Nothing came of his visit.

Gina Kolata, a widely read science writer for the *New York Times*, wrote me:

> You have to ask what newspapers are trying to do. There are hundreds of articles every day crying for out readers' attention. And every article has to tell a reader: 'why am I reading this and why am I reading this now?' The news can feature something incredible, like the solution to Fermat's last theorem or the cloning of a dog, or it can involve some issue that has a big impact on society like health care, or it can be a discovery of something quirky. Newspapers are not there to educate or to teach people about the mathematics that underlies search engines unless there is something you can say about that mathematics that makes it new and compelling. The fact that the mathematics is there is not enough. With most things we use – a car, a iPod, a DVD – most of us don't really care how it works.

Sara Robinson, a science writer who majored in mathematics, and was Writer in Residence at MSRI (Matematical Sciences Research Institute, Berkeley, California) recalled that as a fledgling reporter, she heard from a senior reporter that the goal of science reportage was to give the reader merely "an illusion of understanding of the technical subject matter." She also reported a statement made by Rob Finer, a former editor of the *New York Times*, that

> mathematics has no emotional impact. What physicists do challenges peoples' notion of origins and creations, mathematics doesn't change any fundamental beliefs or what it means to be human.

In view of the fact that the increasing mathematisation of life is changing what it means to be human, this strikes me as a completely imperceptive view of the matter.

NOVELS, PLAYS, AND MOVIES WITH A MATHEMATICAL UNDERLAY

In the last two decades, they have come thick and fast.
- *Morte di un matematico napoletano* (1992). (Death of a Neapolitan Mathematician). A brilliant, eccentric, alcoholic mathematician lives in increasing isolation and commits suicide.
- *Good Will Hunting* (1997). Will Hunting, abused as a child and living a rough life in South Boston and employed as a janitor at MIT, is discovered as a math genius by a Fields Medal winning professor. With the aid of a therapist, his life is then turned around.
- *Arcadia* (1999) by Tom Stoppard. Plot line: A teenage genius living several hundred years ago discovers that entropy of the universe is increasing. Much talk about Fermat's last theorem, chaos determinism à la Newton and Laplace, population dynamics, etc.
- *Fermat's Last Tango* (2000) A Musical. Won an Emmy Award. In fantasy, the spirit of Fermat and of other mathematical greats residing in a heavenly *Jenseits*

or Aftermath, meet up with an obsessed mathematical hero reminiscent of Andrew Wiles of Fermat fame.

- *Uncle Petros and Goldbach's Conjecture* (English translation, 2000) by Apostolos Doxiadis, a mathematician and movie maker. A brilliant mathematician is obsessed by his inability to solve a difficult problem. Giving up, he retires in defeat to playing chess and tending his garden.
- *A Beautiful Mind* (2001). This movie won four Academy Awards. A great mathematician descends into madness and then makes a partial recovery. The emphasis is on the mental state and not on the particular mathematics achieved. It was based vaguely on the biography of mathematician John Nash.
- *Proof* (2005). Won a Pulitzer Prize; also a movie with Gwyneth Paltrow and Anthony Hopkins. A deceased, mentally ill mathematician had a daughter who devoted her life to caring for him. His daughter and a former student of the mathematician come across the proof of an important theorem in the mathematician's Nachlass. The question raised: to whom is the proof due? A second question: has the daughter inherited her father's madness?

The solid impression gained from these stories is that mathematicians are strange and peculiar people.

A DATABASE OF MATHEMATICAL FICTION ETC

Prof. Alex Kasman of the College of Charleston, South Carolina, is a mathematician, an author, and an archivist for literary works with a mathematical underlay. Kasman maintains a website titled *Mathematical Fiction* that contains brief summaries and opinions of about 500 works. I would not have guessed that there have been so many works of this sort and I am indebted to his website for some of my information.

Kasman has divided and cross-listed these works into about 50 categories. Under the rubric of "Medium" we have comic books, films, novels, plays, short stories, TV series or episodes. Under "Genre": Children's Literature, Didactic, Historical Fiction, Humour, Espionage, Fantasy, Horror, Mystery, and Sci-fi. The writers of these works range from satirists such as Jonathan Swift or Stanislaus Lem to novelists like Robert Musil who studied mathematics as a young man. They range from Sci-fi writers such as Jules Verne, playwrights such as Pirandello, or Tom Stoppard (who is a mathematics buff), to humorists such as James Thurber. They include professional mathematicians such as Charles Dodgson, or my doctoral thesis advisor Ralph Boas, Jr.

Kasman's list contains a wide variety of themes, types of works, written by people with different degrees of professional mathematical knowledge and experience. The number of professional mathematicians who write fiction of this type is larger than I imagined. While the whole genre is not easily characterised, I would call about half of the individual works catalogued *sensationalist*. A few examples:

- *The Rose Acacia* by Ralph Boas, Jr. A computer came up with a deal with the devil as to exactly how many terms it takes to arrive at two accurate places of a very slowly convergent infinite series.
- *Pop Quiz* by Alex Kasman. Messages from an alien spacecraft seem to be asking deep questions in algebraic geometry. What is the intent of all the messages?

I could go on and on, citing additional instances of mathematics in fiction. What is clear to me is that there is much material in what I have found that can easily support the assertion that the popular stereotype of the mathematician, as brilliant, somewhat mad, socially inept or reclusive, obsessive, living in the clouds, given to the arcane, the fantastic, is not in error. The mathematicians or mathematics depicted come wrapped in the following sensational themes: magic, codes, espionage, the devil, ghosts, secret messages, other worlds, futurism, madness, autism, apocalyptism, mysticism, the occult, obsessions, prizes, distopias, evil mathematical productions and cults, machines that turn in to sorcerers' apprentices, alternate time concepts. Apparently, there is a steady market for this kind of literature and mathematicians themselves are writers, producers, and readers of it. Mathematics is often regarded by the average person as a kind of magic and this view fits right in with the fictional themes. There is no doubt that the community of professional mathematicians likes to write this kind of material. Pulitzer Prize winning *LA Times* writer Dan Neil was aware of all this when he wrote:

> Never have so many relied on so few to tell us what the hell is going on. Mathematicians have acquired the status of hieratic otherness, a kind of geek priesthood, acting as intermediaries between the unfathomable and familiar. [Definition: geek = an obsessive specialist]

Of course there is another way at regarding these productions. A leading research mathematician told me that he factors out the mathematical descriptions which in any case he finds trivial and concentrates on the story line, e.g., the relation between an uncle and a nephew in the novel *Uncle Petros*.

The average person has had little hands-on experience with mathematics other than, perhaps, doing a few sums and paying interest on a variety of purchases. The common sense view of mathematics and mathematicians – or what masquerades as common sense – is to a considerable extent formed by what the public remembers from grade school or what they now learn from the media. "You do the math" is now a common expression that epitomises distance from the whole magic, incomprehensible, tedious enterprise. It may be that with its abhorrence of mathematics, the general public ignores the mathematics totally and simply goes for a whopping good story. I would suggest a substantial financial grant to an appropriate university department for a survey to test this statement.

IS THERE ANY REALITY TO THESE STEREOTYPES?

I believe there is some and it starts from the simple observation that certain people are good at mathematics and some are not. Some people care about mathematics deeply; they think about it constantly. Other people couldn't care less. Nature and nurture? Presumably, and on the side of nature, common observation leads to the speculation that some people are hard wired for mathematics (whatever that expressions may mean).

If we humans are hard wired, we appear to be not all wired at the same level. At the very far end of creativity, consider Asperger's Syndrome (Hans Asperger, 1906-1980, Austrian pediatrician) about which there has been much speculation. This is a psychiatric condition, a certain type of autism, often characterised by poor social interactions, repetitive behaviour patterns and numerous eccentricities. The mathematical talent displayed by people with this syndrome can be very considerable. Suggested candidates for Asperger among famous mathematicians or physicists have been Isaac Newton, Henry Cavendish, Bernhard Riemann, Stefan Banach, Albert Einstein, Paul Dirac and Srinivasa Ramanujan. Be this as it may. I am sure that most professionals have had contact with some mathematician that they would put in this category. As Clifford Geertz wrote:

> The rational beauties of mathematical proof are guarantees of no mathematician's sanity.

WHERE DOES THE DIFFICULTY OF COMMUNICATION LIE AND WHAT CAN BE DONE?

There is, paradoxically, a disconnection between the substantial mathematisation of everyday life to which everyone is subjected and the extreme reluctance of the general public to learn anything of the subject beyond grade school material. Why should they know more when the relevant numbers and implications are spewed out automatically? (Television star Rosie O'Donnell opinioned: "I think there's no way they should have to teach it [mathematics], now. We have computers.")

Where does the difficulty lie? Does it lie with mathematics education in the lower grades, often taught by teachers for whom the subject is poison? Doesn't every specialised activity from cooking to dentistry have its specialised terms that the public becomes familiar with? Shouldn't grades K-12 provide a basic vocabulary and infrastructure of understanding so that the average newspaper reader will not turn the page rapidly upon encountering the word *mathematics*? Does the difficulty lie with the nature of mathematics which, at its higher reaches, is a difficult subject requiring much more time for absorption and patience on the part of readers than newspaper articles do? Does the difficulty lie therefore with the nature of the media with its promise of instant information and understanding? Does it lie with the journalists who write about recent accomplishments in the field? With the mathematicians themselves who, in the opinion of Dan Neil, constitute a geek priesthood? It would appear that all of the above are operative.

A West Coast mathematical sciences writer (with a BA in the subject) said to me recently that there are now many books written by professionals to popularise their specialties. "And you know what? After five pages, they've lost me."

James Crowley of SIAM points out that

the reason that we tend not to report on mathematics may have something to do with the nature of the discipline itself as well. Because most discoveries in mathematics are incremental, building upon a vast and growing structure of knowledge, there is seldom one breakthrough discovery that can be announced with one big 'aha'!

A number of programmes intended to cure the situation are in place. The professional mathematical societies and a number of individual writers work hard to call attention to a variety of clienteles at every level of sophistication to recent accomplishments in the field. Thus, the American Mathematical Society has a number of series devoted to this task. It maintains a *Math Digest* that summarises popularising articles and articles about mathematics in the popular press. *Mathematical Moments*, "promotes the appreciation of the role of mathematics in science, nature, technology, and human culture" by releasing to schools and colleges attractive glossy flyers, suitable for display, on such individual topics as forecasting the weather, creating better eyeglass lenses, etc. What gets released to the newspapers is information such as the names of recent prize winners, adding to the absolute glut of prize winners of all sorts in today's world.

The Division of Mathematical Sciences at National Science Foundation is trying to gather items about mathematics and its application to use in press releases .This is not an easy job because few people in the mathematical sciences are accustomed to reporting on their work at a level that can appreciated by a general audience. It is not part of their culture. Some institutes such as MSRI in Berkeley have a rotating journalist in residence. The United States Air Force, as a response to fewer and fewer native American students pursuing science, technology and mathematics, has given a grant to the University of Southern California to teach screenwriting to scientists in an attempt to produce movies and television shows that show scientists in more sympathetic ways than what is around. How all these professional projects diffuse into the popular media and public awareness is a matter of conjecture.

WHAT KIND OF REPORTAGE WOULD I LIKE TO SEE?

As an antidote for sensational reportage, I would suggest that newspapers run articles that give a semblance of understanding of the degree to which mathematics that underlies today's world – yes a semblance. We are living in an age which is mathematised and is increasingly so. To realise this, take a look at the front page of a newspaper and count the number of numbers that are on it. Some numbers invoke trivial mathematics, some less so. Go to the business section and do the same. Go to the sports section and do the same. And the mathematics involved is often of long standing and is not material hot off the research pages.

Just think what a different (perhaps better) life we would lead if IQ's were not around to tell us who is "intelligent" and who is not. How would we live if there were no blood pressure or cholesterol numbers to advise us; if there were no pre-election polling on every conceivable issue; if the trajectory of a missile or of a rocket to Mars were computed by pre-Newtonian theories or by guesswork; if, when we went to the supermarket, the checkout clerk took a pencil from behind his ear, marked down the prices of our 18 items and toted them up. Did you know how the House of Representatives is apportioned among the States after a census? There is a mathematical procedure known as the "method of equal proportions" as part of statutory law (Title 2, U.S. Code) and judged constitutional by the U.S. Supreme Court. Have you looked at a weather map with its isobars and isotherms? There are mathematical algorithms that underlie the production of these pictures. What meaning can be ascribed to the statement that "it will rain tomorrow with 40% probability"? Do you know that the diffusion equations have been invoked to advise orchid growers how often to water their plants?

Each of these experiences and hundreds more, have been created by calling in an underlay mathematical ideas and methods, some trivial, some deep and by no means within the abilities of the general public, and as yet few of them have been called to the attention of the public as something worthy of its attention. Coming close to expositions of these mathematical ideas are the periodic releases entitled *Mathematical Moments*, mentioned above. But the average reader cannot handle their vocabulary.

But in contradiction to what Kolata has said, why is it not the duty of a newspaper to educate? Doesn't a paper educate when it admonishes us not to do drugs? When it advises us how to make a tofu soufflé? When it reminds us that there is a congressional election every two years? Or that it is time to push the clock back one hour? I should think that at the very least it should be possible for a paper to educate us to the fact that mathematics is formatting a good portion of today's life and to point out where this is occurring. It need not give the readers a semblance of understanding of the technical mathematics; that is too much to expect. But I should hope that clever writers might point out how mathematics is altering our lifestyles, and do it in a manner that would not lead Garfield the Cat to say "ho hum." Is it too much to hope that quality newspapers might in the future run such articles as:

Professor Hiram Smith shows how eigenvalues help in search engine strategy. President asks Science Advisor what Egg Values are

Numerical algorithms of aero-hydro-elastodynamics used in the design of the Swiss Yacht that won the America's Cup

Mathematical wavelets aid in gallstone treatments at Massachusetts General Hospital

The Surgeon General urges Medical schools to require probability theory for admissions

The Attorney General urges Law Schools to require probability for admissions

Yogi Berra [baseball star and unconscious wit] praises Markoffian applications to baseball strategy

I hope that such items have already been run and that as I read my morning paper with my eggs and coffee I have simply missed them.

POSTSCRIPT

This article represents my view of the American scene vis-a-vis mathematics and the media. My correspondents in Europe tell me that with some slight modifications it describes the European scene equally well.

REFERENCES

Davis, P. J. (2005). Mathematical exhibitions: Reactions and concerns. *SIAM News*, Dec. 2005.

Geertz, C. (1983). *Local knowledge: Further essays in interpretive anthropology.* New York: Basic Books.

Neil, D. (2005). Who knew math was so prime time. *Los Angeles Times Magazine*, Oct. 9, 2005.

Randall, L. (2005). Dangling Particles. *New York Times, Op-Ed,* Sept. 18, 2005.

Robinson, S. (2001). Mathematics and the media: A disconnect and a few fixes. *SIAM News*, Oct. 2001.

http://www.ams.org/mathmedia/mathdigest/

http://www.ams.org/ams/dbis.html

AFFILIATIONS

Philip J. Davis
Brown University,
Providence

ACKNOWLEDGEMENTS

The chapter "Students' foregrounds and the politics of learning obstacles" by Ole Skovsmose is based on his paper "Foregrounds and politics of learning obstacles", published in *For the Learning of Mathematics, 25*. We wish to express our gratitude to the editor-in-chief of *FLM* for authorising the print of the chapter within this volume.

The chapter "Implicit mathematics: Their impact on societal needs and demands" by Yves Chevallard is a reprint from a collection of papers on the theme "The mathematics curriculum: Towards the year 2000: Content, technology, teachers, dynamics" from the 6th International Congress on Mathematical Education, Budapest, Hungary, 1988, edited by J. Malone, H. Burkhardt and C. Keitel. We wish to express our gratitude to the Curtin University of Technology for permitting the reprint of the paper.

NAME INDEX

SUBJECT INDEX

Printed in the United Kingdom
by Lightning Source UK Ltd.
131538UK00001B/64/A